MOVING SAM MALOOF

MOVING SAM MALOOF

4880 Lower Valley Road • Atglen, PA 19310

SAVING AN AMERICAN WOODWORKING
LEGEND'S HOME AND WORKSHOPS

ANN KOVARA, AIA, LEED AP

Other Schiffer Books on Related Subjects:

Esherick, Maloof, and Nakashima: Homes of the Master Wood Artisans, Steven Paul Whitsitt and Tina Skinner, ISBN 978-0-7643-3202-9

Wood in Traditional Architecture, David Campbell, ISBN 978-0-7643-3581-5

California Colonial Homes: Case Studies with Prominent Architects, S. F. "Jerry" Cook III and Tina Skinner, ISBN 978-0-7643-2392-8

Copyright © 2016 by Ann Kovara

Library of Congress Control Number: 2016935658

All rights reserved. No part of this work may be reproduced or used in any form or by any means—graphic, electronic, or mechanical, including photocopying or information storage and retrieval systems—without written permission from the publisher.

The scanning, uploading, and distribution of this book or any part thereof via the Internet or any other means without the permission of the publisher is illegal and punishable by law. Please purchase only authorized editions and do not participate in or encourage the electronic piracy of copyrighted materials.

"Schiffer," "Schiffer Publishing, Ltd.," and the pen and inkwell logo are registered trademarks of Schiffer Publishing, Ltd.

Front cover and book's graphic design concept
by Norah Tahiri
Cover design by Matt Goodman
Front cover images, l to r: Robert Buettner, Norah Tahiri,
Ted Catanzaro, Robert Buettner, Norah Tahiri.
Back cover images, from top, l to r: Norah Tahiri,
Tavo Olmos, Norah Tahiri, Norah Tahiri, Robert Buettner,
Tavo Olmos, Robert Buettner, Tavo Olmos, Norah Tahiri,
Norah Tahiri, Tavo Olmos.
Type set in Sinkin Sans/Minion Pro

ISBN: 978-0-7643-5136-5
Printed in the United States of America

Published by Schiffer Publishing, Ltd.
4880 Lower Valley Road
Atglen, PA 19310
Phone: (610) 593-1777; Fax: (610) 593-2002
E-mail: Info@schifferbooks.com
Web: www.schifferbooks.com

For our complete selection of fine books on this and related subjects, please visit our website at www.schifferbooks.com. You may also write for a free catalog.

Schiffer Publishing's titles are available at special discounts for bulk purchases for sales promotions or premiums. Special editions, including personalized covers, corporate imprints, and excerpts, can be created in large quantities for special needs. For more information, contact the publisher.

We are always looking for people to write books on new and related subjects. If you have an idea for a book, please contact us at proposals@schifferbooks.com.

This book is dedicated to Sam and Alfreda.
Thank you for being.

To remake a thing correctly is to discover its essence.[1]

---

1. Howard Mansfield, *Same Ax Twice: Restoration and Renewal in a Throwaway Age* (Hanover and London: University Press of New England, 2000), 4.

CONTENTS

INTRODUCTION..10
MEETING AT THE SHOP..14
THE LEGEND OF SAM & ALFREDA..22
THOUGHTS ON A VANISHED LANDSCAPE..32
THE SUBJECT IS ARCHITECTURAL STYLE..54
CASE STUDY: MALOOF RELOCATION PROJECT...62
MALOOF NEW RESIDENCE..86
EPILOGUE...94
APPENDIX...96
FURNITURE, ART & LANDSCAPE INVENTORY...101
MAPS, SHOP DRAWINGS & FIELD NOTES...109
PHOTOGRAPHY CREDITS..120
ILLUSTRATION CREDITS...122
BIBLIOGRAPHY..123

INTRODUCTION

Some alterations to our world occur in small and imperceptible ways, but grand modernization schemes impose radical changes. Such programs also generally result in the clash of the old with the new. Examples of vast transformative projects include Baron Haussmann's carving of elegant boulevards through thousands of Parisian housing blocks; Frederick Law Olmsted's creation of New York's Central Park; Daniel Burnham's planning and building of the Chicago 1893 World Columbian Exposition; and the US road building effort that resulted from the Interstate Highway and Defense System Act of 1956.

Major public works projects usually begin with the purchase of land, the relocation of citizens, and ultimately, the demolishing of existing buildings, infrastructure, and landscaping. Common also is the moving of specimen trees as well as historic structures and those associated with a notable person or event. The abandonment of neighborhoods is often the result after the inhabitants are moved out of the construction zone.

Lost along with the physical characteristics of a neighborhood are its social fabric and the personal associations developed by the residents over a period of years. A site becomes the focus of memories and associations; events transpired leave a memory, while the spirit of the original place remains. For instance, a tree in a garden may be special to the owners because it represents a moment shared with their children while planting it. The profound sadness and great joy associated with a property become elements of that site's memory.

> Old things have stories to tell.
> Their stories remain
> A mute voice
> But still we listen.[2]

THE HUMAN SIDE OF HISTORIC PRESERVATION

Large public building programs involve the interaction of people with divergent world views. Differences of opinion regarding basic goals and priorities generate friction between pragmatic engineers and civil servants engaged in the constructing of infrastructure projects, and private individuals fighting to save unique environments, or their own property, from expropriation by the state.

In all grand public schemes some individuals are caught up in the surrounding events like dried leaves in the wind. Such was the case for two octogenarians, master craftsman Sam Maloof and artist Alfreda Louise Ward Maloof, his wife of fifty years. An exuberant orchard of trees, shrubs, and flowers gently framed the Maloof property until the State of California decided that a new highway was needed, and Sam and Alfreda's land was standing in the way.

Sam had built out the Maloof family compound in Alta Loma, California, over a span of forty-five years. His woodworking shop was just as he wanted it. The Arts and Crafts–inspired house he shared

---

2. Valerina Quintana, *This Piece of Earth: Images and Words from Tumamoc Hill* (Tucson, AZ: Tumamoc: People and Habitats, College of Science, University of Arizona, 2013).

with Alfreda was both a peaceful sanctuary and showcase for the woodwork that made Sam famous. Sam and Alfreda celebrated all things "hecho a mano," or made by hand.

Because of the proposed state highway extension, the Maloofs endured an environmental phase that lasted more than ten years, throughout which they fought for their land. As the years went by, Sam and Alfreda felt that their voices were not being heard and they worried about the state employees' ineffective communication with them and their lack of concern for the Maloofs' point of view. Through the eyes of the government entities, the extension of the road was business as usual; they saw progress. For the Maloofs, it was an intensely personal experience, and they feared destruction.

Much was at stake on both sides. On the government's side was the traveling public's need for the completion of a freeway system, job creation, and the expected infusion of federal transportation funds into local coffers. From this point of view, the demolition of the property of some old guy who makes chairs would be for the greater good.

The Maloofs considered the preservation of their mature orchard, creative woodworking shop, and art-filled home to be as valid as the construction of a highway. Sam and Alfreda grew a variety of trees, including California natives and citrus. They placed the highest value on personal expression; to them beauty, inspiration, and craft mattered. All might be lost because of a twenty-mile-long eight-lane expressway. The Maloofs wondered, "Why? And for whom?"

The Maloof family and their attorneys endured a grueling environmental review process that lasted more than ten years as decisions were made regarding the final path of the Route 30/210 freeway between La Verne and San Bernardino, California. One might think that Sam and Alfreda had many years to accept change and adjust to the situation. In reality, the Maloofs were pushed to the limits of their endurance by the pressure of unresolved urban planning decisions.

Finally, Sam and Alfreda's property was deemed historic by the State of California. It was this ruling that saved the Maloof home, main woodworking shop, guesthouse, and some specimen trees from demolition. These elements would be relocated and the remainder of the site demolished under the Maloof Relocation Project, which was an element of the larger freeway extension project. San Bernardino Associated Governments (Sanbag) was the overall administrator for both the freeway extension and the Maloof project. Sanbag is a government coalition that includes the mayors and selected city council representatives from twenty-four cities in San Bernardino County. The State of California Department of Transportation (Caltrans) performed the actual freeway construction work. Burge Construction was the Maloof Relocation Project general contractor.

The final Maloof site historic designation ruling was the result of a stellar collaborative effort among the Maloofs, their attorneys, Sanbag, Caltrans, Federal Highway Administration (FWHA), City of Rancho Cucamonga, and California State Office of Historic Preservation. In the Maloof Foundation newsletter, *The Wooden Latch,* Dr. Hans Kreutzberg explains the saving of the Maloof property this way:

Because the property was deemed eligible for the National Register of Historic Places and federal [transit] funds were involved, it was necessary to enter into a mitigation process under Section 106 of the NHPA [National Historic Preservation Act] of 1966. The mitigation, which saved the Maloof home from demolition, is a model for preserving unique historic structures. According to Dr. Hans Kreutzberg, Supervisor of the Cultural Resources Program for the CA State Office of Historic Preservation, "The Section 106 undertaking on the Sam Maloof / Route 30 Project, was in my judgment, one of the most successful win-win mitigations in CA preservation history. Sam and Alfreda Maloof received fair market value for their historic property, a foundation was established with a three million dollar endowment to perpetuate the woodworking achievements of Sam Maloof, the home was successfully relocated and a new freeway is under construction."[3]

Because the Maloofs lost their family compound, from their vantage point it was not a win but a disaster. Sam and Alfreda did appreciate that part of their world would be saved. According to the State of California's ruling, the contributing factors to the property's historic designation were its "association with a significant person of the past" and "distinctive characteristics of the building by its architecture and construction including having great artistic value or being the work of a master."[4] Of course the master was Sam. The environmental process end result was the relocation of the Maloof home, woodworking studio, and some other significant elements to a new site about three miles up the road. This is the true story of how progress and tradition, public needs and private lives, managed to reach an accord.

---

3. Dr. Hans Kreutzberg as quoted in "Maloof Residence Relocation Completed!!," *The Wooden Latch,* Fall 2002.

4. Bonnie W. Parks and Aaron A. Gallup, "Continuation Sheet 1, California Department of Transportation Architectural Form," Feb. 17, 1989, rev. Jul. 6, 1990.

Opposite: New residence great room containing many of Sam Maloof's iconic furniture pieces.

MEETING AT THE SHOP

Through the open blue doors of master craftsman Sam Maloof's main studio, I entered the red-hot center of American woodworking. Chair patterns hung like bleached bones over the shop windows and filtered the dusty light. Many pieces of beautifully sculpted handcrafted wooden furniture sat waiting for completion in every room. I found Sam standing at his table saw, which was strewn with architectural renderings.

Instead of making furniture, on this day in July 1998 Sam was the host of a meeting in his shop, which I was there to attend. He acknowledged my presence before returning to the discussion he was having with five or six people. Although small in stature, Sam was a strong 82 years old with a direct gaze, a soft voice, and large scarred hands. He wore black-framed glasses, a black T-shirt, jeans, a Navajo silver sand-cast belt buckle, and black cowboy boots.

A man of intense practical energy (to paraphrase Walt Whitman)[5], Sam worked in his shop every day. He loved wood in all its forms. Originally a successful graphic artist, in 1949, at age thirty-three, Sam changed professions to become a woodworker. Self-taught but with talent, hard work, and the loving support of his family, through the medium of wood Sam became a world-renowned designer and builder of both beautiful and comfortable furniture.

A recipient of numerous awards and honors, such as a California Living Treasure nomination and MacArthur fellowship, Sam had by 1998 been a leader of the American Studio Furniture Movement for decades. At the suggestion of industrial designer Henry Dreyfuss, Sam began the production of a chair intended as a gift from himself to President John F. Kennedy. Sadly, it was put away after the President's assassination in November 1963. Sam and his staff did create rocking chairs for two US presidents, Ronald Reagan and Jimmy Carter. Of Sam's work Carter said, "…high tech skill, moral values, and inner spirit all incorporated into a piece of furniture."[6] Sam remains an inspiration to all people not just for the quality of his life's work, his technique, and his success as a full-time woodworker, but also for his ability to express the higher values he found in craftsmanship.

The topic of Sam's meeting at his shop was the design of a new residence for himself and his wife, Alfreda. The couple had celebrated their fiftieth wedding anniversary a month earlier. Sam and Alfreda were in need of a new home for the first time in forty-five years because their family compound was directly in the path of a new eight-lane expressway. The Maloof property at risk was a five-acre working lemon orchard with a 10,000-square-foot house of Sam's own design, a main studio building, finishing shop, guesthouse, old farmhouse, several wood barns, and other outbuildings.

Because the home and main shop had been deemed historic under an agreement with the State of California and therefore were eligible for the National Register of Historic Places, these structures would be saved from demolition. Because the freeway construction effort would take Sam and Alfreda's land down as much as seventeen feet, their property would cease to exist.

The Maloof Relocation Project was created to move Sam and Alfreda's historic buildings to another citrus orchard about three miles away at the top of Carnelian Street in the Alta Loma district of Rancho Cucamonga, California. A small part of the larger freeway effort, the Maloof Relocation Project's objective was to relocate the historic Maloof residence, main woodworking studio, guesthouse, and about twenty mature trees to the new site. The Maloof Relocation Project would replicate the historic context of the existing site at the new location and regain the house and main shop's National Register of Historic Places standing.

---

5. Walt Whitman, "2. Democratic Vistas" (1871), par. 118. "I hail with joy the oceanic, variegated, intense practical energy, the demand for facts, even the business materialism of the current age, our States."

6. President Jimmy Carter, "Sam Maloof Woodworking Legend," Rene Russo, host (Los Angeles: KCET-TV PBS special, 2007).

MEETING AT THE SHOP 17

Chapter opening: Main shop (Building A), chair patterns
Top: West wall of main shop
Bottom Left: Slabs of wood in Maloof Woodworking wood barn
Bottom Right: Roof detail with Maloof logo, new wood barn

After the move to the new site, the Maloofs' historic residence would be reopened as a museum, and it would no longer be their home. For this reason, a new residence would be built for Sam and Alfreda.

The Maloofs' peaceful, art-filled environment was the center of their family's universe. Sam and Alfreda knew that the loss of their private world and the move to the new site would break their hearts. Sam wanted time to stop, and life at their homestead to be just as it was before the road builders showed up; he wanted to win the war for their property. What Sam needed was for the freeway people to consider the Maloofs' desires. During the freeway project environmental phase, Sam had to insist that the public agencies involved with the transportation project recognize that he was a master craftsman and a significant person. He also needed the State of California to admit that the Maloof homestead was a special place. Although he finally obtained this acknowledgment, it was ten years in coming. Sam and Alfreda fought for everything they received.

The year 1998 was important because it was the Maloof family compound's last year as an intact site in its original condition. Construction of the new freeway was scheduled to commence in 1999, and there was no stopping it. Sam and Alfreda's magical place was going to be moved, along with its magic, to a new site. Soon days and nights of unremarkable occurrences at the secluded site would cease to exist. Much would be gained, but many pleasant moments would be lost, such as Alfreda preparing meals for everyone at her stove with produce from their garden, or Sam's daily trips to the Alta Loma post office to pick up the mail and chat with their neighbors. Another small pleasure of Sam's life was his Wednesday lunches with his friends at Walter's restaurant in Claremont, which was owned by his pal, Walt Boldig. Eventually Walt died and the lunches came to an end for Sam and other old friends, such as potters Rupert "Rummy" Deese and Harrison McIntosh.

SAM SPEAKS

Imagine what Sam must have been thinking as he looked at the architect's drawings for their new home spread out on his table saw. Sam knew that after its relocation, the colorful and dynamic art and furniture gallery that they called home would become a static museum. Perhaps he pictured strangers wandering through and around their sanctuary.

The architectural renderings brought to the meeting by the design team were the latest iteration of a new residence scheme that the project architects had been developing over the past six years. The designers began their work with Sam's sketches, which he had provided on the backs of church bulletins or lumber invoices, or on restaurant napkins. The current home design reflected the competing interests of meeting attendees Sam Maloof and Maloof craftsperson Larry White; project architects Jim Wilson and Cary Wilde from Thirtieth Street Architects; Gary Moon, the San Bernardino Associated Governments (Sanbag) freeway project director; and Dave Clark, Sanbag's environmental scientist and Maloof Relocation Project manager.

As envisioned by the architects, the finished house would be a combination of 30th Street Architects and Sam Maloof architectural style elements. Because of the Maloof Relocation Project's limited budget, Moon said it was Sanbag's position that the Maloofs' new residence needed to be a modest utilitarian structure, similar to a caretaker's cottage. After Sam and Alfreda's passing, it was Sanbag's intention that the new residence be used as the Maloof Foundation offices.

When it was Sam's turn to comment, he said that it was his intention, from the beginning, for the new residence to be a Sam Maloof–designed work of art, the house equivalent of a Maloof piece of furniture. Sam outlined for the group what he considered to be the key design elements of the Maloof architectural style, and what he found to be lacking in the current scheme under review before them. For example, under the Sanbag plan, when first constructed the new residence would be missing fine woodworking details

such as hand-built redwood door and window trim, Maloof studio cabinetry, and a spiral staircase.

Sam went on to say that he would like to build the new residence himself so that it would be done right. Every project Sam undertook was carried out to the highest level of craftsmanship. He could not fathom why anyone would want to miss that mark, if it could be achieved. Fortunately Moon did not allow him to build the new residence; he felt it would be too much for Sam. And of course he was right. After the construction of the home was finished, we laughed with Sam about his idea.

During the home design meeting Sam didn't seem to be getting his point across to this audience, so he finally said in a quiet voice, "Well, I think I'll just build a model of it." Larry White took the roll of drawings from the architects, who left without them at the close of the meeting. Larry and Sam later created a redwood model that articulated Sam's vision. In the months that followed, Moon had the architects take the model back to their design studio so that they could study the finished model and redraw their plans in accordance with Sam's design.

ENTER FREDA

The meeting appeared to be over when a door opened and someone entered the shop from the living area of the historic Maloof residence. From the expression on Sam's face I knew this had to be his Freda. She had blue eyes and was dressed in a pink T-shirt and Levis, with her silvery-white hair in a braid down her back, Navajo-style. Alfreda had a light around her, which I came to realize was there all the time. I asked Sam about it once. He said he could see it too, and called it her "ecclesiastical glow."

Alfreda had guessed the meeting was over and had come to invite us into the house for some homemade lemonade. After the noise of the workshop, the house seemed very quiet and smelled like furniture polish. The trees around the perimeter of the residence shaded the windows. Near the doorway from the house into the shop was a small cork-topped pedestal table with four chairs. It was next

Top: Redwood model in demolished finishing shop (Building J)

Historic American Building Survey (HABS) photo, view from front door looking west toward the dining alcove

to a fireplace and some brightly colored walls. Sam told me that he made these pieces in the 1950s. He wanted me to know that Alfreda served a casserole to Rosalynn and President Jimmy Carter at this table, and that the president had two helpings.

THE CHAT IN THE GARDEN

After Moon, Clark, and the architects left (without their drawings), Sam invited me to stay and talk with him. We went out in the garden and sat on some rocks under an oak tree. During our talk Sam was low key, direct, and approachable. I have heard him described as a simple woodworker; I would not say that he was simple. He was, after all, a world-class artist at the top of his game. Apparently there were a few things he thought I should know and that we needed to get straight.

My company had sent me out to the Maloof family compound to attend Sam's meeting in the shop and to be interviewed by Sam for the Maloof Relocation Project construction manager position. My company, DMJM, held the multi-million-dollar construction management contract for the 210-freeway extension. Sanbag was going to cancel that contract unless a construction manager acceptable to Sam was found for the much smaller Maloof Relocation Project. My approval was not viewed as a sure thing because of the eleven people sent out to the Maloof site for an interview; Sam had run off the first ten. I waited for my job interview to start, but instead of talking business, Sam felt like telling me about Alfreda.

Sam was a larger than life character and a physically strong person with tremendous drive, but he had suffered a heart attack in 1990. Alfreda had been in ill health for some time, and at the age of eighty-six she was a few years older than Sam. It was not a sure thing that Sam or Alfreda was going to survive the stress of the move.

Sam told me how worried he was about Alfreda. They had been living with the threat of the freeway running straight through their place for a long time, and it was taking its toll on her. After having already endured many difficult years of uncertainty and litigation as the freeway environmental phase moved inexorably forward, Sam seemed confident that the relocation of the property would end badly for the Maloof family.

At the same time, though, it seemed impossible to Sam that such a special place could be demolished; a place that he and Alfreda had created, and that he had built with his own hands. He was absolutely sure that his family compound would be spared and life would return to normal. Sam said something like, *"After you people are gone, I'm going to put everything back the way it was, so don't change anything so much that I can't put it back."* I assured Sam that I got the message, and I had the feeling I was on Team Sam.

But what Sam really wanted to tell me was how he and Alfreda had first met. It was June of 1947. After completing their World War II military service, both Sam and Alfreda had returned to civilian life. Sam worked as a studio assistant for nationally known painter and muralist Millard Sheets, the head of the Scripps College art department in Claremont, California. Already an artist and having graduated from the University of California at Los Angeles (UCLA), Alfreda was a potential Scripps College master's program art student. On this day both Sam and Alfreda were in the Scripps art building courtyard while students were changing classes. Sam noticed Alfreda while waiting for Millard Sheets outside his classroom. He told me Alfreda was the most beautiful girl he had ever seen. Alfreda walked over to say hello, and to Sam's amazement, they had a brief conversation. Sam said they both went home and told their parents they had met the person they were going to marry. It was more than a year before Sam and Alfreda's first date; they were married a month afterward, in June 1948.

THE LEGEND OF

SAM & ALFREDA

> Previous pages (left): Alfreda Maloof portrait, 1997
> Previous pages (right): Sam Maloof portrait, 1997

Both Sam and Alfreda were the children of immigrants. Alfreda's parents were from Sweden, and Sam's Lebanese parents were among the first Arab-speaking people to arrive in California. Sam was fluent in Arabic and Spanish like his parents, Slimen Nasif Nadir and Anisse Maloof. Upon coming to America, Sam's father used his first name as his last, which is a Middle Eastern custom. Jewish traders in the area translated Slimen as Solomon, and the name stuck. So growing up Sam was known as Sam Solomon.

Sam's father and mother ran a dry goods store in the Chino, California, area during the Great Depression. His mother crocheted and tatted lace that his father peddled. The Maloofs had six girls and two boys: Mary, Victoria (BeBe), Rose, Sarah, Olga, Eva, Sam, and Jack. Sam was the seventh of the eight children and the first son. When he was sixteen, in 1932, the household grew to sixteen people when oldest child Mary George's husband died suddenly; she and her six children moved into her parents' home.

Even as a child Sam loved to draw and to make things out of scrap wood. Because he was born to a large family of simple means, at Sam's birth the destinies of his life were not fixed, but he made brilliant use of the years he was given. Sam attributed his success as a woodworker not only to Alfreda's support but also to the unqualified emotional support he received from his parents, the other members of his large family, and some key teachers. For example, his brother Jack, an elementary school principal who passed away at a young age, helped Sam expand the Maloof family compound on weekends. Together Sam and Jack built walls and framed out room extensions.

SAM'S EARLY CAREER

With the encouragement of his art teachers at Chaffey High School and Chino High School, Sam developed his drawing ability, including mechanical drawing and architectural drafting and design. By the time Sam graduated, he was skilled in calligraphy, cartooning, and caricature. After Sam finished high school he audited some college classes while working full time, but he never graduated from college.

Near the end of his high school experience Sam won a poster contest sponsored by Herman H. Garner, founder and owner of the Vortox Manufacturing Company in Claremont, California. The judge was watercolorist Millard Sheets, who at that time was with the Scripps College art department in Claremont. He was Sam's future employer and mentor. Starting in 1934 Sam worked for Garner as a Vortox art department graphic designer for five years. While at Vortox he designed business press releases, advertisements, technical brochures, and stationary letterheads as well as artwork for the Garners' personal interests. Garner permitted Sam to work shorter hours so that Sam could attend classes, first at Chaffey College and then night school at Frank Wiggins Trade School (today called Los Angeles Trade Technical College, LATTC). Sam took classes there in commercial art, and became quite skilled at advertising design. Because he wanted to make some furniture for his family, Sam enrolled in night school at Chino High School where, besides learning cabinetry and joinery, he obtained experience using their woodworking equipment.

Because Sam said he knew how to use a band saw, he obtained a position as a craftsman for successful Claremont-area industrial designer and artist-craftsman Harold E. Graham. Graham was a friend of Sheets and a part-time instructor in the Scripps College art department. While employed by Graham from late 1939 until he was drafted into the army in October 1941, Sam designed and built window displays for Bullock's department store, display cases, cabinetry, and metal and wood furniture.

Sam's World War II military service was spent in Alaska, where he added surveying and mapmaking to his skill set. Sam used his experience with Garner and Graham and his high school classes in drafting to produce detailed engineering drawings, maps, and large-scale displays. After they returned home from the war both Sam and his brother, Jack, changed their last names from Solomon back to Maloof. Master Sergeant Sam Solomon became Sam Solomon Maloof, private citizen.

By 1946, Sam was thirty years old and working as a graphic designer for a commercial art firm in Los Angeles called Angelus-Pacific. While in their employ Sam created decals and logos and handled other 2-D tasks like cartooning, lettering, color separations, and silk screening. Sam was on his way to a successful career as a commercial artist.

Then Sam had a life-changing thought: to replace the cheap, worn-out furniture in his rented bungalow with pieces of his own design. Because new manufactured furniture was too costly, in his spare time Sam began to create contemporary furniture with clean lines, using borrowed tools as well as school shop equipment. For materials Sam found throwaway lumber, such as discarded sheets of fir plywood previously used for concrete formwork and salvaged red oak floorboards from railroad freight cars. It was Sam's creation of contemporary furniture for his own use in the various places that he lived that ultimately led him to his life's work as a woodworker.

MILLARD HIRES SAM

A 1929 Chouinard Art Institute graduate, Sam's future mentor Millard Sheets was an outgoing, larger-than-life character. As a watercolorist, Millard was the leader of what came to be known as the California School of Watercolor Painting. Although he was not an architect, Millard designed many commercial, retail, residential, and institutional facilities, such as his iconic Home Savings and Loan buildings in Los Angeles. He also created more than a hundred glass mosaic and painted wall mural installations for such major clients as Home Savings and Loan and Bullock's department store. Millard's greatest influence was as the leader of the Scripps College art program and as the director and organizer of major art exhibits held at the Los Angeles County Fair.

After Millard returned home from the war he created some fine personal watercolors that he wished to reproduce. For this reason he needed an excellent silk screen printer. An Angelus-Pacific customer, Millard went looking for Sam when he heard that he was doing serigraphy work at their shop. Millard already knew Sam's background, including his winning high school poster contest entry and his work for Harold E. Graham. Millard hired Sam as a studio assistant in Claremont, and Sam's career took a great leap upward. Not only was Sam back in the Pomona Valley, but now he had the opportunity to work on fine arts projects for California's most prominent artist and visionary art professor.

In addition to silk screening, Sam assisted Millard with a wide variety of tasks in the studio. Sam stretched canvas, made picture frames, and hand-lettered the original artwork for labels, brochures, posters, and exhibition catalogs for Millard as well as for his art dealer, the Dalzell Hatfield Galleries. Sam even accompanied Millard on a painting trip to Mexico in 1946. During the years 1947 to 1948 Sam assisted Sheets with his mural work, including the wall behind the bar in Lawry's La Cienega Restaurant in Los Angeles. Working from small sketches provided by Sheets, Sam developed full-size cartoons of the murals. For this project Sam worked in the middle of the night installing gold leaf, one square at a time.

ALFREDA'S STORY

It was while Sam was working for Millard Sheets that Sam and Alfreda first met, in June 1947. Alfreda was a thirty-six-year-old painter and weaver working on an application for enrollment in the Scripps College art department master's program. She had received her undergraduate degree in education from UCLA, which she attended during the depths of the Great Depression.

In 1935 Alfreda began her association with the Indian Arts and Crafts Board created by the US Congress. She was employed as the director of arts and crafts for the Santa Fe Indian Arts Program. Alfreda also taught Native American school children on several reservations in Arizona, New Mexico, and the Great Northern Plains. During this time she became a close friend and colleague of iconic Pueblo potter Maria Martinez and her family.[7]

Along with Maria, Alfreda was part of the movement to bring back traditional Native American arts and crafts, which had been in decline since the railroads arrived in the late 1800s. For example, at that time Fred Harvey created a series of popular restaurants with gift shops located in railroad depots and other public gathering places throughout the West. Because he provided a major opportunity for reservation craftspeople to sell their traditional handmade work to the traveling public, Harvey impacted indigenous culture. The book *Craft in America* explains this pivotal moment in time for Native American arts and crafts, during which Alfreda and Maria Martinez were people of influence:

Recognizing Indian culture, Congress established in 1935 the Indian Arts and Crafts Board, which had the job of promoting "the economic welfare of the Indian tribes…through the development of Indian arts and crafts and the expansion of the market for the products of Indian art and craftsmanship." Under the leadership of establishing standards for quality, reviving declining art forms, and even creating major exhibitions at the Golden Gate International Exposition in 1939 and MOMA in 1941, to call national attention to the Indians' extensive and expressive talents.[8]

Alfreda resigned from the Indian Service in 1943 to enlist as a WAVE in the Navy. Then in 1946, because her father had recently passed away, Alfreda obtained an early discharge from the Navy to take care of her invalid mother. It was after Alfreda's military tour of duty that she began taking classes at Scripps College, where she met Sam, the love of her life.

After Sam and Alfreda's marriage in June 1948, Sam moved into the modest home with an attached garage that Alfreda shared with her mother in Ontario, California. Alfreda had bought the home for herself and her mother with her G.I. benefits. Sam and Alfreda lived at the Ontario house for almost five years before they bought several acres of land with some ramshackle buildings in Alta Loma, California.

Because Millard demanded that Sam work long evenings and weekends, it wasn't long before Alfreda asked Sam if he married her or Millard Sheets. Among other things, this conversation precipitated Sam's eventual quitting of Millard's employ to go out on his own as a woodworker. While working for Millard, Sam had set up his first woodworking shop in the Ontario house garage and he had obtained a few furniture orders. After the couple's son, Sammy (the future Slimen), was born prematurely in March 1949, Sam and Alfreda agreed it was time for Sam to go out on his own and realize his dreams as a furniture maker.

MALOOF FAMILY COMPOUND, 1953
On Easter Sunday 1953, Sam and Alfreda moved to what would become their Maloof family compound (the original site) with their four-year-old son, Sammy, and Alfreda's mother. Marilou, their adopted daughter, joined the family later that year. (In high school Sammy would change his name to Slimen, his paternal grandfather's name.)

The property on which Sam and Alfreda chose to construct their refuge was on the southeast corner of Amethyst Street and Highland Avenue in the Alta Loma District of Rancho Cucamonga, California, about forty-five miles northeast of Los Angeles. The original site's amenities included a shack, a chicken coop, and an enormous avocado

---

7. Alfreda Ward Maloof, *Recollections of My Time in the Indian Service 1935–1943: Maria Martinez Makes Pottery* (Klamath River, CA: Living Gold Press, 1997).

8. Jo Lauria and Steve Fenton, *Craft in America: Celebrating Two Centuries of Artists and Objects* (New York: Clarkson Potter, 2007), 22, 23.

tree. The Maloof property would expand to its final size after the purchase of several adjacent acres with a farmhouse, which became Slimen's future home.

It was here that Sam and Alfreda were inspired to invent their own nurturing, creative environment. The Maloofs worked together at their homestead for almost five decades. Financial success didn't come to Sam and Alfreda until many years later, but life was good with Sam in the shop and Alfreda in the office, managing Sam's woodworking business. Under Sam and Alfreda's care and as their finances and time allowed, the Maloof family compound grew as Sam added new structures and additional rooms.

At the Highland Avenue property Sam's original woodworking shop area consisted of an old shed and a chicken coop. Sam worked bent over in the coop in the daytime, and stored his tools and furniture there in the evenings. In 1954 Sam poured a new exterior concrete slab.

That concrete slab, open to the weather, served as Sam's shop space for about another year and a half until some friends helped him frame out a room on top of it. Sam's one-room shop was the first element of what was to become master craftsman Sam Maloof's woodworking shop. Under the Maloof Relocation Project this area was moved to the new site as a part of Sam's main shop (Section A).

NATIONAL RECOGNITION COMES TO SAM

With the support of some influential interior designers and home magazine editors, Sam received national recognition early in his woodworking career. For example, in the fall of 1949, Millard's wife Mary suggested to her brother, photographer Harry Baskerville, that the Maloofs might be a good subject for a story. Baskerville interviewed Sam and Alfreda for a *Better Homes and Gardens* article called "Handsome Furniture You Can Build." The article ran in March 1951, and along with the Maloofs' Ontario house living room furniture, it included working drawings prepared by Sam.[9] Sam also provided walnut chairs and bar stools for some of John Entenza's *Arts & Architecture* magazine's high profile Case Study House Program residences in the Los Angeles area under his professional representative, the interior design firm Kneedler-Fauchere. And Sam was on several of the Case Study interior design teams organized by his mentor, Millard Sheets.

Leading West Coast architects, interior designers, and style-conscious homeowners ordered pieces from his one-man workshop. Because of Sam's use of natural materials and his furniture's warm tones, clean lines, and hand-sculpted details, his work complemented the spare, open-plan interiors of mid-century modern Southern California residences. Eventually the world moved on to plastic and metal furniture, but Sam maintained his vision and continued to design and craft wooden pieces the rest of his life.

Sam often repeated what he called "his criteria": beautifully designed, beautifully made. He preferred to make one carefully considered piece of furniture at a time. Sam refused high-dollar offers to mass-produce his furniture, even though his workshop only produced approximately fifty to one hundred works a year. So it is ironic that Sam's first major client was industrial designer Henry Dreyfuss, the creator of some of the most representative symbols of the twentieth-century industrial age. Dreyfuss designs include the black tabletop Bell telephone, the Princess phone, the Polaroid Land camera, the Hoover vacuum cleaner, the 20th Century Limited streamline moderne train, and the John Deere tractor. Sam designed and built more than twenty pieces of furniture for the Dreyfuss Craftsman-style home on Columbia in South Pasadena, California.

Sam's career was on an upward trajectory in 1957 when he was invited to participate in one of the first major exhibitions of studio craft furniture, the *Furniture by Craftsmen* show held at New York's

---

9. John Webster, "Handsome Furniture You Can Build," *Better Homes and Gardens,* March 1951, 258.

American Craft Museum. That year Sam and other kindred spirits dedicated to making things by hand attended the first American Crafts Council (ACC) conference in Asilomar, California. Because woodworking and other crafts were gaining national interest, Sam found himself at the center of the growing American Studio Furniture Movement.

AMERICAN STUDIO FURNITURE MOVEMENT

Along with master craftsmen George Nakashima, Wendell Castle, and Wharton Esherick, Sam is remembered as a leader of this movement. Nakashima, Castle, Esherick, and Maloof represent the first generation of American studio furniture makers. Begun in the 1950s, this collective effort was a revival of the Arts and Crafts Movement associated with William Morris in England. Both movements valued the union between maker, object, and owner. They celebrated studio-created hand-built furniture and eschewed machinery and mass production.

The studio furniture pioneers were alike in their ability to create simple, sculptural forms in wood without applied ornament or historical style. They also shared a common love for the essential warmth and beauty of solid hardwoods, and the belief that architecture and furniture must be functional. This first generation of studio furniture makers continues to influence the next generation, or second wave, even though these post-modernists reject some of the beliefs of the modernists. For this second generation, Sam remains an iconic figure.

Sam freely shared his knowledge and love of craftsmanship on an individual basis in his shop, but also in more formal settings. For example, he taught woodworking through the University of California at Riverside Extension and at the Anderson Ranch in Aspen, Colorado. And although he never had the opportunity to finish college, Sam was highly honored by academia. Some of his prestigious diplomas include his Honorary Doctor of Fine Arts, Rhode Island School of Design (RISD); Honorary Doctor of Fine Arts, Cal State San Bernardino; and Honorary Doctorate of Humane Letters, Aurora University, Aurora, Illinois.

Each piece of Maloof studio furniture is a work of art, created with a particular person in mind. Examples of Sam's life work can be found in the permanent collections of major museums across the country, including Albuquerque Museum; American Craft Museum; Carter Presidential Library; Dallas Museum of Art; De Young Museum; Fuller Craft Museum; Los Angeles County Museum of Art; Metropolitan Museum of Art; Museum of Fine Arts, Boston; National Museum of American Art – Smithsonian; Oakland Museum of California; Philadelphia Museum of Art; Saint Louis Art Museum; The Detroit Institute of Arts; The Huntington Library, Art Collections, and Botanical Gardens; The National Museum – Renwick Gallery; the Toledo Museum of Art; and the White House Permanent Collection.

MALOOF FAMILY COMPOUND 1998

Without the intrusion of the freeway, Sam and Alfreda would have lived out their days at the Maloof family compound with family and friends close by. The Maloofs' son, Slimen, had spent most of his life at the homestead and raised two children there, Aaron and Amy Rebecca Maloof. They had since grown up and moved to town, but they often returned to the homestead. Slimen worked with his father in the Maloof studio until he began working alone in the shop behind his house, which was in the far northwest corner of the Maloof family compound.

Sam's fellow craftsmen at Maloof Woodworking (whom he called "the boys") had been with him for years, including Larry White, Mike Johnson, and David Wade. The team also included Alfreda, who was the Maloof Woodworking business manager, and shop apprentice Kern Briggs. After working in the shop through the move years, their shop apprentice Kern Briggs left the Maloof site for a job as a congressional aide in Washington, DC. Patricio

Casas Rodriguez was in charge of the grounds and his wife, Raquel Lopez, assisted in the house, both at the existing and new sites.

Larry White and his wife, Katherine (Kat), had occupied the Maloof guesthouse since 1992. "They formed a union of mind and spirit that would last for forty years."[10] Both Larry and Kat were artists with studios off site. Larry worked in various mediums, including clay, pen and ink drawings, mixed media, and sculpture. He was a former university-level art teacher and Kat had been his favorite student. She was associated with the Chaffee College Art Gallery, Riverside Art Museum, and dA Center for the Arts in Pomona. A member of the Maloof studio team since 1962, Larry left to open his own woodworking shop in Santa Cruz, California, in 1969. Because Sam was getting older and needed him, Larry rejoined Sam and the Maloof workshop team in 1992.

Maloof craftsman Mike Johnson loved woodworking and was introduced to Sam by a family friend in 1981. Then, to Mike's amazement, Sam offered him a job in the shop. Both Larry and Mike worked on the shaping of the pieces produced by Sam at his band saw. Mike also created miniature copies of Maloof furniture in his spare time.

After graduation from Etiwanda High School in December 1988, future Maloof craftsman David Wade joined the team at the shop. While working part time with Sam, he learned classical woodworking at Fullerton College. After receiving a certificate in cabinet making/millwork and construction technology, David transferred to Cal State University and graduated in 1992 with a BA in crafts. After David's graduation Sam asked him to work full time, and the completion of Maloof furniture became his area of expertise. David worked in the finishing shop that was directly north of the main woodworking shop. On his own time David created beautiful furniture in his own style. He worked in wood and metal, incorporating aircraft parts into his hand-built furniture.

THE REDWOOD MODEL

Sam was a frequent participant in art gallery shows and museum exhibitions, and all of the woodworkers (except Kern) exhibited their work alongside Sam's furniture. Sam's boys were his extended family, and the relocation of the shop and other building elements from the existing site to the new site was stressful for them as well. For example, the Maloof woodworkers prepared the shop for the move and removed the redwood from the demolished finishing shop for reuse on the new finishing shop. Larry, David, and Mike also created the new residence spiral staircase from the cut-offs left over from the Douglas fir trusses that were constructed on site to the east of the house. They bought a B-32 Agazzani band saw to cut the treads. For the new residence window and door trim, the boys used a mix of new clear heart redwood and reclaimed siding from the existing site. Because of all their move-related tasks, the woodworkers had less continuity and time to finish their Maloof furniture projects.

The transition from historic home to museum was bound to be a difficult one for Sam and Alfreda. Sam assumed that the Maloof family would be able to live in both houses, if they were only left in peace. He was excited about the new residence, but for the wrong reasons. Sam assumed that with the help of his attorneys, he could get Sanbag to provide a Maloof new residence that was the house equivalent of a piece of Maloof fine furniture. In his mind the historic residence was the prototype of his Maloof architectural style, and the new residence would serve as a major variation on this theme.

About two months after the meeting at the shop, Larry and Sam finished the design and construction of the six-foot-long redwood model of the new residence. Because Gary Moon had directed the architects to take it back to their design studio and study it, the architects' revised drawings were a good

---

10. Larry White, "Six to Sixty: Life at the Speed of Art—A Retrospective," *The Life Celebration of Katherine White,* March 2012.

representation of the model. When the time came for Larry to present to Alfreda the completed redwood model of their new home, she looked at it and sadly shook her head as if to say, I can't do it. Not so many days later Sam took Alfreda to the hospital. She had been taken ill suddenly with a stomach ailment, and was placed in the hospital overnight for observation. The hospital staff told Sam that Alfreda was not in danger and suggested that he go home to rest.

ALFREDA DEPARTS

It was after midnight on September 23, 1998, when Sam received a call from the hospital. The news couldn't have been worse: Alfreda had died from a massive heart attack. He went over to the guesthouse to ask Larry if he would please tell Slimen his mother had passed away. Sam couldn't do it. Slimen did not answer his phone, so Larry made his way through the grove in the dark to tell Slimen that he had lost his mother.

When I reported for work a couple of hours later, I found Sam wandering in the car parking area. Agitated, confused, and with tears in his eyes, Sam recounted the events of the night. Sam said he deeply regretted leaving Alfreda's side because in so doing he had missed the chance to say a final goodbye to her before she "went away." I don't think Sam ever got over their sudden parting; Alfreda was his guiding light.

Sam, Larry, Mike, David, Slimen, and Kern spent the next three days in the shop putting all of their skill, talent, and love into the creation of Alfreda's Maloof-style casket of walnut, ebony, and zircote. The young men did most of the work because Sam was barely able to concentrate. A very personal celebration of life was held for Alfreda on October 1, 1998, at the Claremont School of Theology's Kresge Chapel, in Claremont, California. The Maloof studio had recently completed the altar furniture for the chapel, including a suspended walnut cross.

After Alfreda's passing Sam designed a Japanese-style Zen shrine for her grave, which was to be their mutual burial plot. As the site Sam chose the sloping hill outside his future master bedroom window. As envisioned by Sam, the entry sequence into this square, low-walled sacred area would be through a tall, post-and-lintel tori gate, which led to a center island surrounded by a koi-filled moat. The Maloof crypt on the island would be topped with heavy natural granite, a 10-foot-by-10-foot 8-inch slab, which he had already bought. Larry built a tiny replica of the Zen garden from Sam's sketches and placed it in the redwood model.

Everything reminded Sam of Alfreda, he missed her so much; but he managed somehow. I assumed it was because he had phenomenal inner resources and a bright inner flame, until I talked to Larry. He said no, not a flame: a nuclear furnace. Yet because the freeway construction project continued to move forward, Sam couldn't stop to grieve. He faced the upcoming changes alone, without his partner and best friend. Only thirty-five days after Alfreda's passing, the deed for the Maloof family compound was transferred from the Sam and Alfreda Maloof Living Trust to Sanbag. And through it all, Sam kept on working. With Alfreda's passing, Sam had also lost his business manager. Bonnie Hannah temporarily filled that position for a few years until Sam hired Roslyn (Ros) Bock. Ros and Sam worked well together, and she is the Maloof Studio business manager today.

Unfortunately, Gary Moon disallowed the Zen burial plot idea and Alfreda's interment in Sam's new backyard because Sanbag technically owned the new site. Moon shared with me a few other things that he was unable to tell Sam, such as that it was inappropriate to bury someone on an active construction site. Moon acknowledged that Alfreda's casket was a work of art. Sam put into that special piece all of his love and a lifetime of skill. For this reason, Sanbag was reluctant to assume responsibility for the casket at an unguarded gravesite at the new site.

View of new residence rooftops from the historic residence's tree house room porch

Unable to bury his Freda as he wished, Sam threatened to call President Bill Clinton at the White House, former President Jimmy Carter, or the Smithsonian. Moon was unmoved. So Alfreda stayed at the mortuary until Sam moved into their new home in May 2000 and the property was turned over to the Maloof Foundation. Alfreda's grave today is covered with wild flowers, the perfect final resting place for a sweet and gentle lady.

The honesty and grace of the years of life that is ending is the real measure of how we die. It is not in the last weeks or days that we compose the message that will be remembered, but in all the decades that preceded them. Who has lived in dignity, dies in dignity.[11]

---

11. Sherwin B. Nuland, *How We Die* (New York: Vintage Books, Random House), 1995.

■ THOUGHTS ON A

VANISHED

LANDSCAPE

Previous pages: Tower room (Section 4)

What did the Maloof family compound on Highland look like? Because of the tightly spaced, wild vegetation along the perimeter, it was impossible to see the buildings from the street. On my first trip to the site, I discovered that the only way in was from a driveway on Highland Avenue, which was full of cars. After parking on a nearby street, I walked back across the road and entered the property through a rolling gate with an *M* on it. Rooted to the spot, I experienced a "Wizard of Oz" sort of moment, like when the world went from black and white to color. Spread out before me was the physical manifestation of Sam and Alfreda's beautiful and meaningful lives. The spell was cast.

The Maloof homestead looked more like a village than like the expected single house with a woodworking shop. Redwood-clad buildings with blue sheet metal pop-up roofs and towers lined the driveway. *Los Angeles Times* reporter Bob Sipchen captured the feeling when he said, "Wheel into the Maloofs' brick and concrete driveway, and Southern California's…yammering boom of hyperbolized banality is left snapping at the gate."[12]

An old lemon eucalyptus tree (*Corymbia citriodora*) windrow lined the north and west sides of the site, and a dense, overgrown working Sunkist lemon grove surrounded the Maloof buildings. The trees growing near the top of the driveway included California live oaks (*Quercus agrifolia*), an 80-foot Deodar cedar (*Cedrus deodara*), and a 70-foot Canary Island date palm (*Phoenix canariensis*). Japanese maple trees (*Acer palmatum*), loquats (*Eriobotrya*), and flowering plants also grew on both sides of the rock-lined drive.

THE OUTER COURTYARD

A car-parking or outer courtyard area at the bottom of the driveway terminated the entry sequence into the Maloof homestead. A California black walnut tree (*Juglans californica*) and several old olives (*Oleo europaea*) near the ends of their life spans contained this area amid low rock walls and concrete cyclopean walls. Frank Lloyd Wright used this type of wall treatment at Taliesin West; he had a lot of rocks, too.

What was a Sam-constructed cyclopean wall? Similar to an arroyo stone wall, it was an unreinforced cast-in-place concrete wall with exposed local stones in natural colors varying from black to brown to green. To begin, a plywood form is constructed, which is filled with rocks and then wet concrete. The stones that are to be exposed on the outside of the wall are wired into place. After the concrete sets up, the wires are cut and the stones are scrubbed clean with a wire brush. Project contractor Bob Buettner thinks Sam built this type of wall as a practical matter because he had a lot of rocks on his property and concrete was expensive.

The cyclopean wall is an element of the Maloof architectural style. For example, it can be seen at the outer courtyard south of the guesthouse, on the main woodworking shop building's west elevation, on the new restroom building, and on the new residence master bedroom's east wall. Sam constructed the outer courtyard cyclopean walls on weekends with his brother, Jack, a school principal who died young in 1966. There was a horizontal seam in the north wall that was south of the guesthouse (Building J) because Sam and Jack stopped pouring concrete into the formwork when Alfreda called them in for lunch. After the brothers returned to their work, because the earlier pour had started to set up, there was a horizontal seam, or "cold joint," in the wall.

Under Bob Buettner's direction, for the Maloof Relocation Project the masonry foreman Jessie Meir and his team built exact replicas of the demolished original site walls at the new site, including the north wall with Sam and Jack's seam.

This sensitive reconstruction of the cyclopean walls at the outer courtyard was a key element in the replication of the historic context at the new site,

---

12. Robert Sipchen, "A Man of the Woods," *Los Angeles Times*, July 24, 1994.

because this area was the former heart of the Maloof family compound, and because of its importance to Sam. The outer courtyard was encircled by the house, main woodworking shop, finishing shop, and guesthouse. The blue doors of the Maloof woodworking shop, where I first met Sam, were in the southwest corner of this area. The gateway into the home's entry courtyard was in the southeast corner.

THE GREATEST AVOCADO TREE OF ALL TIME

The defining natural feature of the original site, a hundred-year-old avocado tree (*Persea americana* "Fuerte"), grew near the guesthouse on the upper west side of the drive. This avocado tree was one of the reasons the Maloofs bought the property. Sam and Alfreda had lived next to the tree in a shack until the guesthouse replaced that building. Under

Top: Cyclopean wall plywood formwork at outer courtyard

Bottom: Cyclopean wall detail: green rock wired to plywood formwork

the Maloofs' care, forty-six years later the tree was 70 feet tall with a 70-foot-wide canopy and a five-foot-diameter trunk. Sam liked to say that it was the largest avocado tree in the state of California.

This avocado tree was the mother of dozens of other avocado trees on the property. In Sam's mind it was a historic monument, equal to a giant sequoia or redwood tree. He even concludes the story of his life, *Sam Maloof, Woodworker*, with a picture of his avocado tree.[13] Sam considered the big avocado tree significant enough to be worthy of preservation, and expected that its presence would save the Maloof homestead from the freeway bulldozers headed their way. Unfortunately, to others it was just another old tree. In the words of William Blake,

The tree which moves some to tears of joy is in the eyes of others only a green thing that stands in the way.[14]

Because of its age and size, it was obvious to all but Sam that this avocado tree would be impossible to relocate to the new site. Fortunately, Sam did not insist on watching as it was cut down; he did ask to have some of it. Sam was told that because the Maloofs had sold the property, all of the trees (living or dead) were State of California property. In the end, the tree was unceremoniously demolished to accommodate the move of the guesthouse through the site and out on to Amethyst Street.

NAMING OF THE STRUCTURES

All of the buildings at the original site were given identifying letters during an early freeway project survey.[15] The structures on the site at that time and their freeway project names include the historic Maloof residence and office (Buildings A–G); historic main woodworking shop (Buildings H–I); finishing shop/display area (Building J); Alfreda's storage/west wood barn building (Building K); the guesthouse (Building L); wood barn (Building M); and

---

13. Sam Maloof, *Sam Maloof, Woodworker* (Tokyo and New York: Kodansha International, 1983), 216.

14. William Blake, "Letter to the Reverend John Trusler," Aug. 23, 1799, in *The Letters of William Blake* (1906).

15. Bonnie W. Parks and Aaron A. Gallup, Continuation Sheet 1, California Department of Transportation Architectural Form, Feb. 17, 1989, rev. July 6, 1990.

Top and Bottom: At Sam's request, the remains of the great avocado were moved to the new site

the garage/wood barn (Building N). During the Maloof Relocation Project we used these identifying letters to describe the buildings. After the structures were relocated to the new site, the City of Rancho Cucamonga Fire Department continued to use this labeling system.

GUESTHOUSE (BUILDING L)

The guesthouse porch was southwest of the huge avocado tree. In 1961 Sam and Alfreda built the guesthouse after demolishing their original home, the shack. The guesthouse is a square building with a covered deck on all four sides. Mid-century modern white glass globe pendant light fixtures hang from the rafters and encircle the porch. A pyramidal hip roof or cupola with a handmade weather vane sits on a tarpaper and asphalt shingle roof.

This structure is clad in 1-inch-by-12-inch unfinished redwood vertical planks, which is the exterior siding theme used throughout the site; it contributes to the sense of an integrated whole. The 4-inch-by-4-inch porch posts support the deck and the porch roof while resting atop rocks from the site, in the Japanese manner. The posts were connected with raised, rosette-shaped malleable cast-iron washers. Sam used this type of washer throughout the site for heavy wood timber bolting on posts and trusses. Characteristic of Sam's material palette also is the dark navy stain used on the porch rafters and posts.

The interior of the building consists of one room with a bath and adjacent kitchen. Maloof craftsman Larry White and his wife, Kat, lived in the guesthouse during the Maloof Relocation Project years. At that time it had cork walls, a wood stove with a blue tile surround, and a Maloof-style brick floor. The guesthouse was too new to qualify for a historic designation, but it was considered a contributing element to the Maloof historic site. So, along with the house and shop, the guesthouse was partially disassembled, moved, and reassembled at the new site in the same position as it held at the original site.

HABS photo, guesthouse NE corner before move

Top: HABS photo, guesthouse west elevation with Larry White's garden
Bottom: HABS interior photo, guesthouse (Larry and Katherine White's home)

In typical Maloof fashion, before the relocation of the guesthouse to the new site, the wooden plank porch contained an eclectic mix of objects, such as a row of old wooden theater seats, an antique apothecary's cabinet, and a citrus grove smudge pot. As of this writing, the theater seats are still on the porch, and the smudge pot is across the garden near the Maloof Foundation office.

GARAGE / WOOD BARN (BUILDING N)

On the east side of the drive across from the big avocado tree and the guesthouse were two utilitarian structures: garage/wood barn (Building N) and adjacent wood barn (Building M). Both buildings were clad in the typical Maloof unfinished redwood siding. Sam built Building N in 1984; it had several blue sheet-metal shed roofs with various slopes and heights as well as an open tower with a school bell in it. The school bell tower was constructed during the Maloof site-building boom that occurred as a result of Sam's winning the MacArthur Fellowship grant in 1985.

When viewed from the top of the driveway, the roof configurations of the garage/wood barn (Building N) gave it the appearance of several buildings; hence the feeling that one had encountered a village. The overall effect was one of exuberance. Because both garage/wood barn (Building N) and wood barn (Building M) were not contributing elements to the historic nature of the site, they were demolished and not replicated at the new site. So unfortunately, at the new site, the view from the top of the drive does not include that of the original garage/wood barn (Building N).

During Sam's building phase at the original site, sheet metal roofs were considered by the City of Rancho Cucamonga to be a utilitarian or industrial building material and were not permitted for residential use. Because of the secluded nature of the property, the City allowed the use of Sam's blue metal roofs at the

HABS photos, the Maloofs' garage / wood barn (Building N)

Maloof original site until the move of the historic residence to the more exposed new site precipitated a lot of discussion, newspaper articles, and the subsequent changing of the City code.

SAM RELATES TO WOOD

Both garage/wood barn (Building N) and wood barn (Building M) held part of Maloof Woodworking's hardwood collection. Sam's hardwood was stored in at least three other major woodsheds, as well as various other locations around the site. Altogether at that time, Sam had about 200,000 board feet of lumber.

At the original site the lumber storage areas included mainly fine hardwoods for use in handcrafted furniture making, which Sam had been collecting for almost five decades. Many types and grades of wood were represented because Sam's lumber collection evolved over a long period of time. Some of the hardwood was air-drying in the storage sheds, but most of the lumber was already steam- and kiln-dried.

The Maloof hardwood collection included California walnut, English walnut, English brown oak, white oak, curly maple, tiger's eye maple, fiddleback maple, cherry, old growth teak, and other fine woods. The exotic colored woods represented included purpleheart, bubinga, zircote, Burma padauk, Macasser ebony, Gabon ebony, and Indian and Brazilian rosewood; ivory was never used, for any reason. Of the exotics, Sam used only zircote and rosewood in the construction of furniture. With its dramatic dark brown and light cream tones, zircote became a favorite of the Maloof studio. The other exotics were used in a decorative manner for utilitarian purposes, such as plugs over screw holes, or as butterflies, pulls, pegs, and inlays.

Some of the woods found in the storage areas had been harvested in the Maloof family compound grove, including lemon, walnut, eucalyptus, and avocado. Walnut was used in furniture construction, and lemon was used to accent the forms of the pieces. Occasionally, Sam found a new use for trees blown down in storms. One such example is the peeled avocado log hanging below the ceiling in the historic Maloof residence's tree house room. Sam explained his relationship with wood this way:

Some woodworkers talk about the necessity of contemplating a piece of wood and letting it tell them how it wants to be used. This is fine, but time is precious. Personally, there are so many pieces of furniture for which I have mental drawings and there are so many more pieces of wood in my future that I have no time for leisurely conversations with a single piece. My communications with wood, therefore, are very efficiently condensed. I have had so many conversations that I now use a sort of shorthand. I relate intensely to wood. The pieces that will become furniture are chosen with a mixture of common sense and love, and there is no reason for this process to be long and arduous. Though I have a continuous love affair with wood, there are other things in my life that are much more important; among them are my family and my friends.[16]

WOOD BARN (BUILDING M)

Wood barn (Building M) was between and close to both Sam and Alfreda's garage (Building N) and the north end of the house (Building A-G). Building M was the original garage. It was a tall narrow building, well suited for the vertical storage of large planks of wood. It had two shed roofs with asphalt rolled roofing, and a rolling barn door. Sam's tractor and other farm equipment were stored outside it in the grove to the east.

HISTORIC MALOOF RESIDENCE (BUILDING A–G)

The historic residence was a 10,000-square-foot, U-shaped structure that served the Maloofs as a combination home, office, art gallery, and furniture showroom. Just like Sam's furniture, the home was finished on all sides, but because vegetation grew

---

16. Sam Maloof, *Sam Maloof, Woodworker* (Tokyo and New York: Kodansha International, 1983), 65.

THOUGHTS ON A VANISHED LANDSCAPE 41

New wood barn building, Maloof Woodworking dining chairs

right up to the edge of the structure, only portions of the house could be seen at any one time. Many first-time visitors went on a long journey around the site looking for the main entrance into the home.

A rustic double redwood gate in the southeast corner of the outer courtyard led to the home's entry courtyard and a large Japanese maple tree (*Acer palmatum*). Across the entry courtyard from the gate, posts with malleable iron bridge washers held up the entry portico. Sam crafted a series of wavy vertical redwood slats set on edge for the front door.

Occasionally a visitor would ask what the design of the door represented, or what the designs of his other gates, doors, latches, and hinges represented. This puzzled Sam. He always responded that his piece was meant to be an abstract design and was not representative of anything.

Top: Handcrafted entry gates to historic residence entry courtyard

The front door opened into the dining/living room; to walk into the museum today is to enter the Maloof home because, of course, they are the same. The home's interior reflected Sam and Alfreda's united spirits and vision, including the design aesthetic, flow of the rooms, and the arrangement of the art collection. The end result was a masterful use of materials, objects, light, color, and space.

Together Sam and Alfreda spent their entire lives thinking about, collecting, and creating art, which is one of the reasons they kept adding on to their house. They needed to provide more space for their growing collection of furniture and arts and crafts. The house also served as Sam's showroom for the display of his furniture. He created vignettes or environments to illustrate how the individual pieces were to be used. As noted by Huntington Museum curator Harold B. Nelson, "Within the house that Sam built there exists today one of the finest collections of post-war arts and crafts in Southern California."[17]

Notable was the Maloof furniture collection's complete integration with the overall character of the interior and exterior of the historic residence. For example, all of the doors, cabinets, latches, hinges, and gates, as well as window and door trim areas, were handcrafted out of redwood by the woodworking studio team in the same simple, elegant style as the furniture. The softly rounded redwood millwork had Maloof-style expressed dovetail joints. For the purposes of the move, the Maloof Art and Furniture Collection was insured for ten million dollars.

MALOOF DINING / LIVING ROOM

The first thing one notices upon entering Sam and Alfreda's living/dining room is the iconic 1970 walnut eight-drawer rectangular dining table to the left, sitting under a Karl Jennings forged-iron chandelier. A bright yellow built-in couch is to the

---

17. Harold B. Nelson, *The House that Sam Built: Sam Maloof and Art in the Pomona Valley, 1945–1985* (San Marino, CA: Huntington Library, Art Collections, and Botanical Gardens, 2011)..

right, next to an orange wall and the fireplace. Straight ahead, through the wooden cabinets, one can see into the kitchen.

Imagine Alfreda at her stove, wearing down the bricks; or imagine, across from the kitchen in the alcove to the right, Sam coming in from the shop through the blue door. Behind the kitchen to the west, Sam added a breakfast nook. Numerous spider plants hung from the ceiling from a pulley system that Sam rigged up to help Alfreda with the watering.

ALFREDA'S OFFICE

Alfreda's office was a few steps from the kitchen and shop doors. Built in 1957 as a carport, the office was in the oldest part of the house. This room had a sliding glass door facing the outer courtyard. Next to that was a baby grand piano; Sam used the top of it as a layout area for important documents. A Sam Maloof desk was in the center of the room. The cabinets on the east wall held Maloof Woodworking records, Christmas cards, art collection purchases, personal letters, and other memorabilia dating back to 1953. An antique cast-iron wood-burning stove sat in front of the south wall, which was masonry. A rectangular dormer with clerestory windows was added above the center of the room in 1989.

Outside the office, next to the fireplace and over a cork-topped table, hung a Millard Sheets painting called *Convoy to India*. With a thin, white brush Millard signed it: "From Millard and Mary to Sam and Alfreda on their wedding day." This is an example of how every room in the house was a personal, art-filled space.

The rooms on the first level of the house also had raw common bricks that were uneven in size. The bricks were placed without sand over black roofing felt paper that was on top of the concrete floor slab. Without sand and because the bricks were of different sizes, they made a clinking sound when walked upon (and they still do). Sam liked to say that the bricks would pinch your toes if you walked on them barefoot.

THE SITTING ROOM

Because of the U-shaped floor plan, as one journeyed through the house it would mysteriously unfold. Around to the left of the Maloof dining room table was the hallway into the children's bedroom wing. And through the etched glass French doors next to the dining room table was a small sitting room. Straight ahead, an antique cast-iron wood-burning stove sat on a blue tiled pedestal in front of a blue tiled wall, which was next to the upper level staircase and a closet.

During the move years, this closet was full of bolts of orange, hot pink, and yellow Jack Lenor Larsen wool upholstery fabric; he was a friend of Sam's. The upholstery fabric had been in the closet so long that it became vintage. The boys used the last of the orange on a 1950s Maloof-style couch they made for the 2011 *The House that Sam Built* exhibition held at the Huntington Library, Art Collections, and Botanical Gardens in San Marino, California.

The sitting room, the adjacent second dining room/alcove, and a full bath were added in 1958. Then in 1977–78 Sam cut a hole in the roof above the sitting room to build the second story balcony/mezzanine area and a straight-run staircase. A full-size Phillip Green mahogany canoe that Sam installed still hangs in the center of the space. The mezzanine included a Maloof furniture display area. A cozy tree house room and second-floor art gallery were added to the house in the mid-1980s.

In this part of the house and in many other rooms, Sam and Alfreda installed ceiling fans along with sliders and jalousie-type operable windows to enhance the breezes. For heat the Maloofs used, in many rooms, antique cast-iron wood stoves set on concrete pads with decorative blue tile below and behind. The mature orchard provided shade from the hot sun, but because of the trees, the house was dimly lit. So Sam and Alfreda added various types of lofts, galleries, light monitors, clerestories, pop-ups, towers, and gable roofs to bring in more natural light. On many of these roof forms Sam installed distinctive Maloof-blue standing seam sheet metal roofing.

Long ago Sam obtained large white frosted glass pendant incandescent light fixtures with black steel powder-coated frames from the shipyard at the Port of Long Beach. Some of these were installed in the second dining/alcove room, which was the next area experienced after the sitting room. The alcove contained Alfreda's paintings representing Southwest Native Americans. Sam had some extra shipyard light fixtures in storage that he chose to install in the great room of the new residence, along with a few others that came from Slimen's vacation cabin near the Klamath River in Northern California.

TOWER ROOM AREA

Traveling east from the second dining/alcove room, the next rooms encountered were Marilou's bedroom and the tower room. Before the move Marilou's bedroom was used as a storage area; during the move phase we found in there a very old and quite large green beaded African hat with a matching tunic. Before the historic residence relocation, Marilou's bedroom had a leaky pop-up roof with clerestory windows. Burge Construction repaired the water-damaged areas before the move of the structure. Repaired also were the water leaks in the breakfast nook area, pyramid/guestroom, and several other places.

In the late 1980s the hallway outside of Marilou's bedroom was transformed into a tower room with two circa-1912 unpainted carousel horses. Because it is a small room with a high ceiling, visitors sometimes ask what possessed Sam to build such an odd space. Ros Bock tells the story this way: A building official at city plan check told Sam that building permits were required only for rooms greater than a certain size. Sam's response was to build a room with a footprint under the permit limit, which is why the tower room is a 10-foot-by-10-foot two-story room with a pyramidal roof and clerestory windows.

Top: HABS photo, Maloof office, looking south toward Sam's desk

Bottom: HABS photo, balcony/mezzanine area looking southwest toward gallery

The tower room acts as the knuckle or joint between the east/west and north/south wings of the house. It would also serve as the interface between the public and private areas, except that most of the house was public space. Alfreda and Sam were comfortable with tours through their bedroom. The only private rooms in the entire house were the master bathroom, Alfreda's bathroom, the children's bedrooms, the office, and the closets.

MASTER BEDROOM WING

The master bedroom wing contained the master bedroom, master bathroom, and Alfreda's sitting room. Sam and Alfreda's bed was in the middle of the room, topped by a Navajo blanket and a Sue Hertel ceramic dog. A Maloof horizontal walnut dresser was on the south wall behind it. A large Mexican tree of life sat on a table in the center of the west wall. This clay sculpture was the only object broken in the move of the art collection, and, fortunately, it was easily repaired. A gable clerestory pop-up roof lit the center of the roof. Larry White told me that Sam added this type of architectural treatment to the master bedroom and Marilou's room before a tour of the grounds by the American Institute of Architects (AIA) Inland Empire chapter in 1993.

A white-painted partial-height shelving unit and room divider held Sam and Alfreda's pre-Columbian pottery, as well as Native American folk art, shoes, belts, and jewelry. This shelving unit acted as a room divider between the master bedroom and Alfreda's sitting room, which was next to the door into the spiral staircase room. Alfreda's space had a daybed, a child's rocking chair used by grandson Aaron, a vintage toy collection, and Native American paintings made by friends and students.

Top: HABS photo, upper art gallery looking northeast

Bottom: HABS photo, master bedroom looking north toward the Maloofs' pre-Columbian pottery collection

SPIRAL STAIRCASE ROOM

Because the Maloofs had many guests, the historic residence had places to sleep tucked into various corners. For example, upon stepping through the door into the spiral staircase room from Alfreda's sitting room, a twin bed was against the wall under a large Sue Hertel painting of a horse. Alfreda's bathroom was to the right of this area, and the spiral staircase was to the left.

Sam built the spiral staircase up to the sleeping loft with Larry White, Mike Johnson, and David Wade. It is made of apitong wood, and the story is that it required 100 clamps to hold it together during assembly. Under the stairs was the Maloof family awards area.

Formerly known as Alfreda's studio, the spiral staircase room had an exterior door that opened out to the entry courtyard between the Japanese maple tree and outer courtyard gate. Sam built a stained-glass transom above the door that spelled out the word "Alfreda" in multi-colored glass. Before the disassembly of the house, even while the art and furniture was being packed up, Sam actively rearranged and added to the collection. For example, Sam brought home from a show and installed next to the transom door a life-sized plaster sculpture of a skateboarder. The artist modeled the piece after himself. Sam enjoyed the piece and the fact that it was made by a young artist. It required one of the largest crates of all.

Between the spiral staircase room and the pyramid/guest room (the last room in the house), Sam and his team built a glass-enclosed covered bridge with wooden deck flooring in 1984. A ceiling-hung Indonesian rice goddess sculpture with outstretched arms provided a welcoming presence.

Opposite: HABS photo, Maloof historic residence, iconic spiral staircase viewed from above

PYRAMID / GUEST ROOM

A former carport, this room derived its name from the pyramidal clerestory roof that was added to the center of room in the late 1980s. The pyramid/guestroom was adjacent to the guest bath and contained twin beds, a wooden desk held up by one peeled log set atop a rock, and a glass window box. This window display case held Sam and Alfreda's fine collection of Pueblo pottery. Because world-renowned San Ildefonso potter Maria Martinez was a close friend of Alfreda's, the Maloof collection contained about thirteen examples of her work.

An exterior door to the east led to a small porch with a fig tree (*Ficus carica*), both of which were moved to the new site. A Maloof-style door on the west side of the room opened into the outer courtyard,

near where the journey through the house began. Back out in the sunshine, across the outer courtyard and to the left, was the door to the main woodworking studio (Building H-1).

MAIN WOODWORKING STUDIO (BUILDING H-1)

Before the move, the main woodworking studio (Building H-1) consisted of one building with four distinct areas. The portions of the building included Sam's shop and the west storeroom; Larry and Mike's shop; a wood barn shed accessed from Larry and Mike's shop; and the dust collection room.

The largest room was along the south side of the building. It was called Sam's shop because it included his famous band saw, joiner, and surface planer. The east wall of this room shared a common wall with Alfreda's office in the historic Maloof residence.

Larry White and Mike Johnson's shop was in the northwest corner of Building H-1. Their workspace had a cyclopean partial-height wall and a rolling barn door on the west side that led to a galvanized cattle tank filled with water plants, fish, and a frog. Black-and-white poster-size photos of woodworkers Slimen and Paul Vincente taken in the 1960s covered the east and west walls. These posters and other ephemera, such as flyers from past shows, were carefully removed before the relocation of the structure and then reinstalled in the same locations at the new site. Mike's workstation was on the west side of the room, and Larry's was on the east. The south wall had an old cabinet system; this, too, was reconstructed at the new site. Through the exterior door on the north wall one went through an exterior passageway to reach the finishing shop. The passageway included Larry's bamboo collection.

> Opposite, Top: HABS photo, guest bath
> Bottom: HABS photo, pyramid room/guestroom looking southeast toward the Maloofs' Pueblo pottery collection
> Right: HABS photo, main workshop before the move

SAM WORKS

Each piece of Maloof furniture was produced with a particular client in mind. The joinery, ebony plugs, wooden butterflies, and other details would be fully integrated into the design.

Some believe that Sam created each piece of Maloof studio furniture alone. (In consideration of her father, woodworker George Nakashima, Mira Nakashima calls this "the myth of the lone craftsman."[18]) Sam and his woodworkers were able to work together to realize the vision and to sustain each other in the achievement of the highest level of craftsmanship; together they created world-class furniture.

So in tune was he with the creative process that, for Sam, the act of woodworking was a transcendental, holy experience. Sam's emotional connection to the beauty of wood was the progenitor of his furniture design and production. The results of his life's work celebrate this natural material.

Sam selected the wood for each piece from the hardwood collection in his wood barn. Sam would use one of the many patterns hanging on pegs over the windows, but most often he would lay out his new scheme as a full-size chalk drawing on the concrete shop floor in front of his band saw. After finding what he was looking for in the grain of the wood, Sam would roughly cut out the parts, such as the legs, on the band saw. The cut pieces of wood were assembled on the floor until the furniture elements were ready to be clamped together. When he was satisfied with his creation, he would hand it over to his woodworkers. Sam provided comments and guidance while his woodworkers handled the final shaping, gluing, and finishing process. Maloof woodworkers Larry White, Mike Johnson, and David Wade shaped and finished every piece.

It is well known that Sam would cut out the pieces free-form on his band saw with the safety gate up. Sam was able to accomplish this dangerous task unscathed (most of the time) because of his ability to work with the concentration of a Zen master. Photographs of his hands tell the real story. One day while my assistant, Dolores Moreira, and I were working in our construction trailer, we heard someone yell, "He's done it again!" After his woodworkers took him away for emergency treatment, we ran into the shop and I found part of his thumb near his band saw.

---

18. Mira Nakashima, *Nature, Form, and Spirit: The Life and Legacy of George Nakashima* (New York: Harry N. Abrams, Inc., 2003).

FINISHING SHOP (BUILDING J)

Sam and his staff built the finishing and display shop (Building J) during the years 1988–1989. Building J was southwest of the guesthouse and directly north of the main woodworking studio. At the original and new sites the woodworkers called this building "David's Shop," after Maloof craftsman David Wade; David calls it Building J. Before the move a long rectangular room that served as the finishing shop was on the west end of the building, and a holding area for completed pieces was on the east end, in the same position as the Maloof office at the new site. In actuality, the historic residence was the showroom.

Maloof Woodworking: Evans chair in progress

Building J was not eligible for historic designation because it was a new building. Still, it was replicated because it was a contributing element to the historic context. So a new structure similar to the former Building J was constructed at the new site. The existing redwood siding was removed carefully from the original building, brought up to the Carnelian Avenue property, and installed on the newly constructed Building J. Back at the original site, the remains of Building J were demolished to make way for the freeway.

OTHER BUILDINGS

After Sam and Alfreda learned that the freeway might wipe out their homestead, Sam and his associates kept building at the existing site. Under these circumstances Sam, his associates, and his family members added several new rooms to the historic residence, including the pyramid/guest room with its attached glass-enclosed breezeway, and a guest bath. Sam and his team also built the finishing shop (Building J) and the wood shed known as Mike's shop.

Mike's shop was the large, freestanding, unheated wood barn building directly west of Building J. The woodshed does not have a letter because it was constructed after the completion of the initial freeway project property survey of the Maloof family compound. Because of this the anonymous building was not replicated at the new site.

Built in the 1960s, Building K was a small wooden storage building located west of the historic residence and south of the Maloof studio. Sam called it "Freda's Building." The east side of it was used for general storage of family goods like the Christmas ornaments and Alfreda's art supplies. The west side held special lumber, such as planks of Macasser ebony wrapped in newspaper, and valuable planks of maple. Mike's shop, Freda's building (Building K), and Slimen's house were deemed to be non-contributors to the historic context. So after the last section of the historic residence went up the road to the new site, the freeway builders demolished all of these structures along with the finishing shop (Building J), wood barn (Building M), and garage/wood barn (Building N).

MOVING SLIMEN

Slimen lived in a circa-1940 farmhouse with a woodworking shop and greenhouse. His place was on the southeast corner of Highland Avenue and Amethyst Street, in the far northwest corner of the property. At the new site, this is the location of the visitor's parking lot.

Recognizing that their days as a family at the homestead were coming to an end, in 2000 Sam bought a ranch house with a barn and two old Arabian horses for Slimen in Mentone, California. Slimen was reluctant to move because he had lived at the Maloof homestead his entire life. Yet Slimen had to go because his house was right on top of the new Amethyst Bridge construction zone. Amethyst was the single remaining Route 210-freeway project bridge to be built, and the start of construction for this structure could not be delayed. So the last activity on the Maloof Relocation Project schedule at the original site was to move Slimen's personal possessions about thirty miles to his new ranch in Mentone.

The morning after the last section of the historic residence had departed for the new site, through the original site's Highland gate we saw bare ground instead of the Maloof family compound. The final remnants of Sam and Alfreda's world included a strip of vegetation along Highland, and Slimen's house. He was still in it.

The freeway builders parked a D-9 Caterpillar earth moving machine in Slimen's yard because demolition of his farmhouse and the Amethyst Bridge construction were going to occur the next day. Dolores and I had been up all night moving the two-story historic residence dining room section (Section 2), but as Slimen needed to vacate the premises for all time, after breakfast we went over to his place with the Graebel movers and two moving vans. When Slimen opened his front door I said, "I am sorry, Slimen, but today is your day to move."

Slimen had not packed anything, so the Graebel team took over and quietly handled the situation. Atthowe Fine Arts movers had already taken the important "Slimen collection" of Maloof furniture, including an iconic Evans chair, for Sam's upcoming major retrospective, *The Furniture of Sam Maloof*. This exhibition was held from September 14, 2001, through January 20, 2002, at the Smithsonian American Art Museum's Renwick Gallery in Washington, DC.

After the movers began their work, Slimen left for Mentone with his friend, Sioux Bally, to await the arrival of the moving vans. When Slimen's house was completely empty, and the movers had departed for Mentone, Dolores and I locked Slimen's front door and turned off the lights at the Maloof family compound for the last time.

Top: HABS photo, north side of guesthouse, porch with theater seats

Bottom: HABS interior photo, guesthouse living area with cork walls and large skylight

THOUGHTS ON A VANISHED LANDSCAPE 53

Top: HABS photo, looking north at passageway between the finishing shop and Mike's (future) shop

Bottom: HABS photo, west elevation of main woodworking shop; finishing shop at left

THE SUBJECT IS ARCHITECTURAL STYLE

> Previous pages, Left: New residence, north patio area
>
> Previous pages, Right: Relocated historic residence, east elevation

Besides iconic furniture and the Maloof family compound, Sam created a distinctive and cohesive architectural style over his nearly five decades at the original site. The built environment at the Maloof homestead felt like an integrated whole because of Sam's self-created building treatment and consistent material palette. Maloof standard details are found in all of Sam's buildings in varying degrees. Both the historic residence and Sam's new residence express all of them.

The exterior building treatment applied consistently by Sam included 1-inch by 12-inch-wide untreated, overlapping, rough-sawn redwood vertical siding; natural stone and concrete arroyo or cyclopean walls; sloping standing seam blue sheet metal roofs; shed roofs, gable roofs, and pyramidal-roof towers; dark navy-blue stained exposed heavy timber trusses with malleable cast-iron bridge washers; and wrapped wooden posts with malleable iron washers set on rocks.

Some of the elements most reminiscent of his hand-built furniture design work are handcrafted clear heart redwood doors, and window and door trim with expressed dovetail joints. Walnut was used for latches, hinges, and cabinets. Sam's craftspeople oiled and hand rubbed all of the hand-built millwork with the "Maloof finish," which consists of one-third linseed oil, one-third raw tung oil, and one-third urethane varnish, semi-gloss.[19]

Sam used common brick floors set on black paper without sand, bull-nosed whitewashed exposed pine ceiling boards, pendant "shipyard" lights of black powder coated steel with white frosted glass, antique cast-iron wood-burning stoves, and Maloof-crafted art glass windows.

Characteristic of Sam and Alfreda's unified approach to design was their use of light and space to create dramatic areas and their consideration of the house as art gallery and furniture showroom. Sam's furniture was staged along with artwork and museum quality decorative objects to create interesting and discrete environments. These staged environments allowed potential clients to visualize how Maloof furniture could be used.

A few examples of the Maloofs' approach to interior design includes second-floor galleries with wooden railings draped with Navajo blankets and rugs; ceilings hung with sculptural elements like canoes, flying rice goddesses, and forged iron chandeliers; glass window display cases with ceramics and other art pieces; and spiral staircases. In a magazine article on the unique Maloof environment, writer Walt Harrington asked,

The question everyone wants answered, would Sam Maloof's genius have blossomed if he had not first created this world in which to live and work? In other words, did his genius create this place, or did this place create his genius?[20]

INFLUENCES

Sam and Alfreda's life experiences, artistic bent, social attitudes, and talent informed Sam's furniture work, the Maloof built environment, and Sam's architectural style. His early career experiences and later work with his mentor, Millard Sheets, also contributed to his development as a craftsman. Millard influenced Sam's aesthetic sense and shaped his world view because he introduced Sam to the world of art, with its different mediums, such as ceramics, weaving, painting, and sculpture, as well as the artists working in those mediums in the local Claremont art colony. In particular, Millard's belief that art should be fully integrated with life had a profound effect on Sam.

Sam's direct experience with the energy of the 1950s mid-century modern movement was an important influence on his work. For example, some Maloof pieces were included as part of the décor in the Los Angeles–area Case Study House Program

---

19. Sam Maloof, *Sam Maloof, Woodworker* (Tokyo and New York: Kodansha International, 1983), 68.

20. Walt Harrington, "An American Craftsman," *This Old House*, March/April 1998.

designed by cutting-edge architects such as Charles Eames, Richard Neutra, Eero Saarinen, and Pierre Koenig. Working in this contemporary aesthetic must have contributed to Sam's use of straightforward and clean lines. Case Study House Program architect Donald Hensman was a frequent visitor to Sam's place during the Maloof Relocation Project years. The two remained close friends until Hensman passed away in 2002.

Other local influences on Sam, his furniture production, and his building program included the Pomona Valley with its mountains, arroyos, citrus orchards, and beautiful hazy light, as well as the Craftsman-style architecture created by local Arts and Crafts architects Charles and Henry Greene. Located near the Arroyo Seco in Pasadena, California, the Greene brothers' Craftsman (or American Bungalow) style of architecture is a part of the American Arts and Crafts Movement, which itself is an offshoot of the English Arts and Crafts Movement founded by William Morris. Some of the masterworks of Charles and Henry include the Gamble, Blacker, Thorson, Pratt, and Freeman Ford houses.

Sam was a longtime friend of Randell Makinson, the architect responsible for the restoration of the Gamble House (now a museum), and who served as its first director. Makinson was subsequently named as a founding member of the Maloof Foundation Board of Directors, and Sam served as an active member of Makinson's group, Friends of the Gamble House. In September 2002, the Friends of the Gamble House honored Sam with the second Gamble House Master Craftsman Award for Excellence.

The award ceremony was held in the garden of the Blacker House, which is in the Oak Knoll area of Pasadena. A Craftsman-style masterpiece, the Blacker House reflects the Greene brothers' strong interest in Japanese vernacular architecture and furnishings.[21] Sam and Alfreda Maloof seem to have been impressed with the Japanese culture, as shown by their arrangement of structures at the original site as separate pavilions. As an American Craft Council trustee, in 1978 Sam attended the World Craft Council meeting in Kyoto, Japan, along with Alfreda. While in Japan, the Maloofs visited sacred Japanese sites and viewed examples of vernacular architecture.

Regarding architectural theory, some designers have been able to successfully articulate their style and building programs both in words and in the built environment. This list includes the Greene brothers, Alberti, A. W. Pugin, William Morris, Frank Lloyd Wright, Le Corbusier, and Mies van der Rohe. However, this kind of intellectual approach to design never took place in Sam Maloof's mind. The Maloof family compound was the creation of a group of people, led by Sam, using their heads, hands, and hearts. The buildings, landscape, and hardscape at the Maloof site reflect their social situation, materials available at the time, their high level of craftsmanship, and individual talent.

TWENTIETH-CENTURY ICONS OF DESIGN

Randell Makinson writes in his first book about the Greene brothers, "What is far more important was… the idea that architects could design and build furniture expressly related to their interior designs, a concept that was already being expounded in the International Studio magazine by the English Arts and Crafts architects."[22] Concerning this subject, Charles Greene remarked: "It is impossible to describe the harmony that may be obtained when the furniture and fittings are all designed with the house."[23]

Besides the Greene brothers, many other architects have created iconic furniture designs in the style of their own unique architecture, including Charles Rennie Mackintosh, Frank Lloyd Wright,

---

21. Randell L. Makinson, *Greene & Greene: The Passion and the Legacy* (Layton, UT: Gibbs Smith, 1998), 93.

22. Randell L. Makinson, *Greene & Greene: Architecture as Fine Art* (Salt Lake City: Peregrine Smith Books, 1977), 61.

23. Charles S. Greene, "Bungalows," *The Western Architect*, July 1908.

Ludwig Mies van der Rohe, Charlotte Perriand, Le Corbusier, Charles Eames, Florence Knoll Bassett, Eileen Gray, George Nelson, Eriel Saarinen, Arne Jacobsen, and George Nakashima. Absent from this list are twentieth-century furniture and architectural designers Ray Eames and Isamu Noguchi, because Eames was a modernist painter by training and Noguchi was a landscape architect.

FURNITURE MAKERS DESIGN BUILDINGS

Conversely, a few master woodworkers have developed their own unique architectural style, one that expresses their woodworking themes and that is reminiscent of their hand-built furniture. For example, with the historic Maloof residence and the new residence, Sam Maloof proves that a master furniture maker can design and produce architecture that directly relates to his furniture designs. This is one of the most important results of the Maloof Relocation Project.[24]

Swedish furniture designer Greta Magnusson Grossman is another example of a master woodworker with a specific architectural style. Her residential designs were reminiscent of her hand-built furniture design work. Magnusson Grossman's mid-century modern Hurley House in Los Angeles, California, is an expression of her architectural style. Besides influential interiors in the 1930s, master woodworker Wharton Esherick also designed his home in a style reminiscent of his furniture. His friend, architectural great Louis Kahn, designed Wharton's studio.

HISTORIC ARTISTS' HOMES AND STUDIOS

Along with Georgia O'Keefe, Winslow Homer, Frederic E. Church, The Pollack-Krasner Studio, and the Wyeths, both Sam Maloof's and Wharton Esherick's unique homes are on the National Trust for Historic Preservation's list of historic artists' homes and studios. Sam and Alfreda's place was the thirtieth property to be added to this list. The description of the Maloof property on that register includes the following:

One of nine children of Lebanese parents, who immigrated to the United States, Maloof was born and raised in Chino, California. Maloof began handcrafting the residence he shared with his family in 1954, using whatever materials were available to him at the time. Over the years, he created a home of unique beauty and artistry that is the setting for his furniture and for the extensive art collections he gathered with his wife.[25]

---

24. Tina Skinner and Steven P. Whitsitt, *Esherick, Maloof, and Nakashima: Homes of The Master Wood Artisans* (Atglen, PA: Schiffer Publishing, Ltd., 2009).

25. National Trust for Historic Preservation, "Historic Artists' Homes and Studios Program (HAHS): Sam Maloof Historic Residence and Woodworking Studio," 2003.

Opposite: Maloof eucalyptus burl table in new residence great room

Opposite: New residence spiral staircase looking southwest toward blue front door

Top Left: New residence round loft window with tree branches over front door

Top Right: A Sam Maloof handcrafted picture frame

Bottom: Iconic Sam Maloof dining table with chairs in historic residence

CASE STUDY: MALOOF RELOCATION PROJECT

Sam and Alfreda lost the battle to keep their homestead, but gained a new site at the top of a steep road near the mountains, 5131 Carnelian Street, Alta Loma, California, which is part of the City of Rancho Cucamonga. The grandeurs of the new site and its defining natural features include panoramic views across the Pomona Valley, mountain vistas, and a steep ravine, or arroyo.

When the trucks with the relocated Maloof building sections first arrived at the new site, it was approximately five and a half acres of sloped, wide-open old citrus orchard with exposure to the neighbors on all sides. The original Carnelian Street landscaping included orange, tangerine, grapefruit, and native Californian oak trees, along with continuous eucalyptus windrows to the north and east.

The natural landform of the arroyo occupied the undeveloped eastern edge, along with the eucalyptus windrow and some existing wildlife habitats. One day during construction we heard the birds screaming near the arroyo. We ran over to see what the fuss was about, and saw a bobcat passing through.

SCHEDULES, CODES, AND STANDARDS

We had to manage three intertwined schedules: the Route 210 freeway extension construction schedule, the Maloof Relocation Project schedule, and the activities of Sam's life. Unfortunately, it was not possible to operate on "Sam time" and move at a pace that was comfortable for him because the Maloof Relocation Project was on the critical path of the freeway project. That meant that any major delay in the relocation project schedule would cause a slip in the projected opening day for the entire freeway project. Gary Moon at Sanbag gave us as much time as he could, but in the end, it was not enough time for Sam to adjust to the situation.

Before the move of the property, a comprehensive Historic American Building Survey (HABS) recording project was performed to create a historic record that would document the built environment at the original site. The HABS survey was used for the National Register of Historic Places application process, as an aid in the replication of the Maloof family compound, and as a tool for the original site's architectural and industrial heritage documentation for the Library of Congress.[26]

Contractor Bob Buettner prepared a digital photographic record of the built environment for the Maloof Relocation Project before the demolition of the Maloof original site by the freeway builders.

BASIC INVENTORY GUIDELINES

Under the terms of the relocation agreement with Sanbag, Sam and Alfreda agreed that their historic residence would become a museum, and that this new institution would contain the majority of the Maloof art and furniture collection; this plan took until 2013 to execute by the Maloof Foundation. The Maloof estate took years to settle because Sam furnished the new residence with some of the elements from the historic Maloof residence, so when he passed away, the task of clarifying personal property from museum collection was left undone.

Fortunately, prior to the relocation of anything in 1999, teams were formed to inventory the elements of the Maloof family's lives. For example, Sam's craftspeople documented the Maloof studio's woodworking shop heavy equipment, including Sam's famous band saw, several lathes, table saw, surface planer, etc. All of the objects included on the various lists eventually came back to the new site on Carnelian except for Slimen's furniture collection and the pieces loaned by Sam's customers for the *Furniture of Sam Maloof* show at the Smithsonian.

MALOOF ART AND FURNITURE MOVING CATALOG

Sam selected two artists, Katherine White and his granddaughter, Amy Rebecca Maloof, to catalog Sam and Alfreda's collections. The end result was the forty-page-long "Maloof Art and Furniture

Left: New site dedication, on his birthday, January 24, 2000: Sam Maloof signs concrete sidewalk west of the main shop

26. US Department of the Interior, "Historic American Buildings Survey: Sam and Alfreda Maloof Compound" (San Francisco: National Park Service, Oct. 16, 2000).

Moving Catalog," an excerpt of which is included in the appendix. The catalog contained the art objects' "tombstone" data, such as name, title, artist, the general description, and overall condition of the piece prior to moving. Color photos of the art objects were taped to index cards. This inventory was completed on a room-by-room basis with a copy provided to Cooke's Crating and to Sanbag.[27]

SMITHSONIAN-BOUND FURNITURE

While we were taking his shop and house apart, Sam was selecting pieces for his September 2001 retrospective at the Smithsonian's Renwick Gallery.[28] Many of these pieces had been taken to earlier shows and exhibitions. Some other key Maloof studio furniture had never been off the site previous to the move to the new site, including Sam and Alfreda's dining room table. The farthest these works had ever traveled was from Sam's woodworking shop to the attached historic Maloof residence.

Jeremy Adamson, US Library of Congress director of collections and services in Washington, DC, organized the Renwick show and was present in Alta Loma to coordinate the exhibition furniture selection with Sam. Adamson conducted an inventory and oversaw the crating and transportation of the selected furniture pieces. The items were labeled by Cooke's Crating and transported to Cooke's warehouse until the time came for the Atthowe movers or Action Air to transport the furniture to Washington, DC. The final Smithsonian-bound furniture catalog item list as prepared by Adamson is included in the appendix.

LANDSCAPE ELEMENT INVENTORY

The project landscape architect, Woody Dike, and his team created the Landscape Element Relocation Inventory, which included the major trees, smaller plants, and exterior art. This inventory was a part of the historic record of the site and was a useful tool for the reconstruction of the original context. We referred to this list during the construction phase because the vegetation in the original large and wild garden was removed before we could start recreating the historic context at the new site.

EXISTING SITE PREPARATION

The building sections were to be moved through a new gate onto Amethyst Street because of the curves in Highland Avenue between the Maloof main gate and Amethyst. The demolition of the grove was to start with the creation of the on-site portion of the haul route needed to bring the disassembled historic building sections out onto Amethyst Street.

Sam loved wood, but he also loved trees. One day while Sam was in a meeting in his shop with Los Angeles art gallery owner Toby Moss and her husband, Sanbag sent a crew with a bulldozer to remove all of the trees west of the house and main shop.

When the heavy equipment operator started his work, Sam ran after the bulldozer, yelling to the operator to stop. Failing in that effort, he started yelling that he wanted to salvage some of the wood. The machine operator could not see or hear him and we thought Sam was going to be run over. David Clark, Sanbag's project manager, asked the heavy equipment operator to stop, and then ruled that Sam could not have any of the wood.

In a short period of time, the entire grove was demolished from the west side of the buildings to Amethyst Street. The dense shade at the house had been replaced with hot light and dust, and it was possible to see the road from the kitchen window. Sam's beloved trees were now debris pile elements with their roots in the air. Sam felt like he had lost a loved one. Many people at the site cried that day.

Several days later, Sanbag demolished the trees immediately next to the home and graded the area smooth. Up to this point it had been impossible to

---

27. Amy Maloof and Katherine White, "Maloof Art and Furniture Moving Catalog," June 12, 2000.

28. Jeremy Adamson, *The Furniture of Sam Maloof* (Washington, DC: Smithsonian American Art Museum, Renwick Gallery, 2001).

obtain a comprehensive view of any building elevation because the vegetation grew up to the perimeter of the house. After the experience with the grading of the haul path, Bob Buettner and I were worried that Sam was not going to be able to handle the demolition of the vegetation next to the historic residence.

Bob and I brought Sam over to see the home that was now surrounded with just dirt. With one of us on each side of him, we were ready for anything, but Sam didn't move or speak; he was rooted to the spot. When he could speak, Sam said it was beautiful, and that it looked like a village. He walked around the house, admiring it. Larry said he often would see Sam, near the end of his life, sitting in the new residence great room, just looking up at the historic Maloof residence on the hill above the new house.

NEW SITE PREPARATION

While the landscape demolition, haul route grading, and historic structures disassembly were occurring at the existing site, Burge Construction proceeded with the utility relocations and rough grading work at the Carnelian site. The changes made to the new site terrain were significant. Approximately 3,000 cubic yards (CY) of soil was cut from the site, 4,500 CY of fill that had been taken from the site was added, and 2,400 CY of imported fill was brought in from elsewhere.

Moving lots of dirt in a citrus orchard results in the loss of trees, but Sam assumed it would be possible to change a sloped site into a flat one without removing any of the trees, or altering the terrain. Tangerine, lemon, grapefruit, orange, as well as the other trees to be demolished were shown on the City plan check–approved construction drawings. For some reason, though, one of Sam's favorites at

Top: Historic residence east elevation after demolition of landscaping and rough grading

Looking south toward the new residence and main shop east end, with survey stakes for historic residence in foreground

the new site was missing from the drawings, a huge oak tree (*Quercus agrifolia*) that was in the middle of the future visitor's parking area.

So on the first day of site preparation at the new site, Sam and Slimen stopped the work after this oak was chopped down. A long period of arm waving and yelling ensued. It ended when Sam quietly said to me, "Well, all right. Just don't do it again."

But, of course, more trees disappeared. Then, after most of the eucalyptus trees in the windrow along Almond Street were removed, Sam insisted I talk with former State of California Historic Preservation Officer (SHPO) and Maloof director Knox Mellon about the damage caused by the team and what we were going to do about it. Protective orange snow fence was installed around all of the trees that were to remain, and Bob Buettner worked with the earthwork subcontractor to ensure that the rough grading would be handled in a more sensitive manner.

At this point Gary Moon banned Sam from the new site. Sam felt humiliated. Sam's safety was a concern to Sanbag because he was, after all, an octogenarian. Later, after the main woodworking shop was brought up to the Carnelian property, Sam's movements were restricted to a fenced-in area close to the buildings. Back at the original site, it was challenging keeping Sam out of the construction zone during the disassembly of the woodworking shop and guesthouse.

BUILDING DISASSEMBLY STRATEGIES

Because the main woodworking shop and Maloof home were deemed historic and worthy of preservation, Sanbag funded the cost of their move to the new site. In the initial discussions with the government regarding the status of their property, Sam and Alfreda were responsible for the move of the guesthouse or any other non-historic structure. Sanbag moved the guesthouse anyway under the final terms of the Maloof agreement.

In general, how the buildings were disassembled corresponds to how the structures were constructed by Sam and his team. How did Sam react to the deconstruction of his home and shop? Apparently Sam told his business manager, Ros Bock, that he was worried we were going to find out that he was a terrible carpenter. Maybe Sam was right; the construction of the two-story wall between the dining room and the sitting room was an issue.

John Kariotis, a structural engineer with decades of historic preservation experience, prepared the master building disassembly plan. Brad Sutton, American Heavy Moving and Rigging's project manager, working with team leader Bob Buettner, developed the disassembly shop drawings that were submitted to the design team for approval. I used the John Obed Curtis book as a guide to understand the subject of historic structure relocation.[29]

Bob and Brad's shop drawings contained significant differences from the bid documents. All departures from the structural-engineer-of-record's plan were based on the Burge and American teams' forensic analysis of the as-built condition of the buildings and on the American team's structure moving experience. This was acceptable because under the traditional design-bid-build construction contract, the sequence of the work is part of the contractor's means and methods.

American Movers cut the gypsum wallboard (drywall) at three feet above the floor in every room of the house and shop. Horizontal steel I-beams (move beams) were bolted to the wooden wall studs below the cut. Then the move beams were bolted in place across the room from one wall to another. The building sections were supported in the vertical direction by wooden cribbing that was stacked from the top of the move beams to the ceiling. Large openings, such as the main shop's east end, were stabilized with 2-by-10-by-10 X-bracing.

The American team installed a series of synchronized hydraulic jacks under the move beams. As controlled by Brad Sutton, the jacks worked together to lift the move beams with the attached building elements so that they were free of the concrete slab. After the move team attached rigging to each building section, Brad used the jacks and rigging to guide the building sections onto flatbed trucks.

THE SHOP COMES APART

The main woodworking shop, the first structure disassembled, was cut into four sections. Sam's personal work area with his band saw, which occupied the entire south side of the building, was called Section A. Larry and Mike's shop area, which encompassed the northwest corner, was deemed Section B. Section C was the wood barn room that was connected to the northeast corner of Larry and Mike's shop. The dust collection room attached to the north side of Sam's shop area was called Section D.

Project structural engineer John Kariotis's woodworking shop disassembly plans differ from the shop drawings submitted by Burge Construction and the American move team. In the original structural engineering construction drawings, Larry and Mike's shop (Section B) was to be moved along with just the west end of Sam's shop area (Section A), and the remainder of Sam's shop area would travel by itself. However, Bob Buettner and Brad Sutton disagreed with Kariotis regarding the proper location for the historic shop building's separation joints. The final configuration, as created by the move team, makes more effective use of the bearing walls. Their scheme uses the bearing wall between Sam's part of the woodshop (Section A) and Larry and Mike's shop area (Section B) as a separation joint. For this reason, the bearing wall between Sam's work area and Larry and Mike's shop area moved with their shop (Section B). Bob Buettner explains the technical reasons for this:

---

29. John Obed Curtis, *Moving Historic Buildings* (Washington, DC: US Department of the Interior, Heritage Conservation and Recreation Service, Technical Preservation Services Division, 1979).

68 MOVING SAM MALOOF

The revised separation between A & B was a natural separation line because Section B was constructed perpendicular to A and needed the least amount of disturbance to the framing and finishes to separate. Also, the revised Section A was more stable to crib and transport keeping that long section together.[30]

Much discussion occurred between the structural design engineer, Bob, and Brad regarding the move of the common wall between the east end of Sam's shop area (Section A East) and the west wall of the office, which was actually in the historic residence. In the end, this common wall traveled with the Maloof office portion of the historic residence (Section 1). The wide-open eastern end of the Sam's shop area (Section A) was stabilized with X-bracing and protected from the weather with tarps.

GUESTHOUSE DISASSEMBLY

Of course, everyone hoped that the historic building fabric could withstand the stress of the move without disintegrating during the process. Burge Construction corrected some structural deficiencies to ensure the stability of the buildings. For instance, before the guesthouse could be moved, the Burge construction team replaced many of the termite-eaten exterior four-by-four porch posts and rafter tails. Because the guesthouse was a raised structure on posts, this structure had to be moved in one piece along with its porch. And the move of the guesthouse off of the existing site was complicated by the fact that it had a basement.

After disassembly, on American's truck the guesthouse was forty feet wide, which is the width of the neighborhood streets from curb to curb along the historic Maloof structure haul route. Along the entire route, each neighbor agreed to have their mailbox at the curb rotated or moved before each of the moves. The neighbors also graciously consented to have any tree branches overhanging the street cut to twenty

---

30. Robert Buettner, e-mail to Ann Kovara, Sept. 2013.

CASE STUDY: MALOOF RELOCATION PROJECT 69

Opposite: Raising of the dining room/kitchen (Section 2) onto move beams at original site

Top: Dining room/kitchen and balcony/mezzanine (Sections 2 & 3) on move dollies

Bottom Left: Balcony/mezzanine northwest corner (Section 3) on move dollies

Bottom Right: Balcony/mezzanine northeast corner (Section 3) with tree house room on move beams

feet above the curb, if required. This was in fact necessary because of the height of the Maloof building towers and other vertical building elements.

Even after the weather vane and cupola were removed the guesthouse was still two stories tall, so overhead utility line relocations were required along the haul route. Before the first move, the utility companies undergrounded many of the overhead utilities. For each subsequent move the telephone company sent out a technician with a special pole to lift up any remaining low-hanging wires.

The guesthouse cupola, handcrafted weather vane, entry platform, and steps were removed and stored. Inside the guesthouse, some attached elements had to be prepared for the move, including the cork walls, ceramic tile wall behind the cast-iron stove, and the brick floor.

DECONSTRUCTING THE HISTORIC RESIDENCE

The historic Maloof residence disassembly began after Sam had moved to the new residence, the finishing shop (Building J) was reconstructed, and the main shop and guesthouse reassembly were completed. The Maloof home was disassembled into seven sections during the relocation process, according to the following plan.

The historic residence office (Section 1) contained the Maloof Woodworking office, an exterior closet, and the entry alcove between the home and the shop's east end. The dining room/kitchen (Section 2) was two stories high; the first floor rooms included the dining room, kitchen, breakfast nook, west alcove, former children's bedroom/laundry room, and a bathroom. Section 2's upper floor included the second floor professional art gallery. The balcony/mezzanine (Section 3) was a two-story element, with a first-floor sitting room, second dining room/alcove, bathroom, porch, and staircase; upstairs were the second-story balcony/mezzanine, tree house room, and furniture display area.

The tower room (Section 4) area encompassed Marilou's bedroom, the two-story carousel horse tower room, and the master bathroom. The master bedroom (Section 5) had the master bedroom, Alfreda's sitting room, and a closet, as well as the pre-Columbian pottery, Indian clothing, and toy collections. A two-story space, the spiral staircase (Section 6) first floor, included the spiral staircase room with the family awards corner under the stairs, and Alfreda's bathroom. Up the spiral staircase was a sleeping loft. The pyramid room (Section 7) included the guest bedroom, or so-called pyramid room, the guest bath, Pueblo pottery window display case, exterior landscape shed, and a porch next to an old fig tree. Section 7 also included the enclosed bridge between the pyramid room area and the spiral staircase (Section 6).

EXISTING CONDITIONS

The Maloof home's physical condition was difficult to assess initially because the grove came right up to the edge of the building. Before the demolition of the trees, it was impossible to view more than a portion of any building elevation; it wasn't immediately obvious that the work included the disassembly, move, and reassembly of old wooden structures containing dry rot, termite damage, and some code violations. For example, existing roof leaks needed to be addressed and fixed by the contractor before the structures could be disassembled, including Marilou's bedroom's clerestory windows, the breakfast nook east wall, and the guest bedroom ceiling. Large tarps provided weather protection for the structures during the disassembly and reassembly phases of the work.

Another feature of the original Maloof homestead was its dangerous and out-of-date building systems, such as the electrical wiring. In the course of disassembly, Bob Buettner sent an electrician over to take a look at the electrical panel under the stairs to the upstairs gallery in the historic residence. The electrician ran back to tell Bob that it was surprising that the historic structures hadn't burned down.

Because pack rats had been gnawing on the electrical insulation covering the wiring in the house and attached shop, the insulation was completely gone. It wasn't a potentially dangerous situation; it was one that was happening now. Bob shut down the power to the house and shop until the wiring could be replaced. Sam commented, "Everything was working fine five minutes ago."

NUMBER THE BRICKS!

During the disassembly phase, Katherine White and Bob Buettner noticed a definite wear pattern in the bricks throughout the house, including in front of Alfreda's stove. They put forward the idea that numbering the bricks would be the right thing to do. The Burge team removed and numbered each brick before crating and storing them until needed in the reassembly phase. After the historic residence was moved to the new site, the bricks were put back exactly as Sam had installed them.

SPIRAL STAIRCASE

The American team moved Sam's iconic spiral staircase with the remainder of the spiral staircase room (Section 6) while still attached to the ceiling. Cooke's Crating protected the staircase in place with packing material before disassembly of the building structure. Then the art piece was lifted up and separated from the foundation along with that portion of the house.

MOVE SEQUENCE PLANNING

The Historic Structures Move Sequence Plan is still posted on the wall in the museum's water heater room outside the breakfast nook. The overall effort to move Sam included the relocation of twelve building sections on eleven trucks in ten moves. There were only eleven flatbed trucks because two small rooms belonging to the main woodworking shop were on the same vehicle, including the wood barn room (Section C) and dust collection room (Section D).

All of the historic building moves took place at night except for main woodworking shop Sections B, C, and D. These small elements of the shop were moved at one time in a daylight move. Larry and Mike's area of the shop (Section B) was the first section to go over the curb into Amethyst Street; it waited there for a truck with the other two sections. This second truck, which preceded Section B up the Carnelian hill, included the wood barn room (Section C) and the dust collection room (Section D).

In the second move, Sam's favorite work area with his band saw (Section A) was relocated in one piece on one truck, and because of its size, was the first night move. Disconnected from its surroundings, at the new site the workshop looked pretty rough, but it still had the same personality. When Sam's friend Randall Mackinson came up to the new site, he remarked that it looked like the old shop, "and we even moved the cobwebs."

Larry and Kat's guesthouse was the third move. Because of the building element's width and height it was relocated as one section on one truck. Sam and Alfreda's home was the subject of moves four through ten; it required one truck for each of seven sections. The specific order of the house moves was:

- Office (Section 1)
- Pyramid room (Section 7)
- Spiral staircase room (Section 6)
- Master bedroom (Section 5)
- Tower room (Section 4)
- Balcony/mezzanine (Section 3)
- Dining room/kitchen (Section 2)

The structure movers carefully planned the order of the move and placement of the building sections. Like puzzle pieces, the sections were moved in the order they were "plugged in" at the new site. The structure movers identified six sections that were key to making the arrangement work. All four sections of the main woodworking shop were installed first, followed by the guesthouse, and finally the office (Section 1) of the historic residence. Upon arrival at the new site, the remaining building sections were grouped around the six key sections in a U-shaped pattern.

The American team used a system of rigging and hydraulic jacks to pull the guesthouse and all of the historic residence sections along the forty-eight-foot-wide path between the south side of the main woodworking shop and the top of the steep slope leading down to the new residence. Brad Sutton guided the sections into final position using rigging and the synchronized jacks.

MOVE ROUTE LOGISTICS

All of the night moves of the Maloof historic building elements started around 11 p.m. The sections were moved in the middle of the night because after

disassembly, most of the building sections were as wide as the street (forty feet). We walked alongside the trucks all of the way, even in the cold winter rain. As all this was physically and emotionally too much for him, Sam waited up at his new residence for the parts of his buildings to arrive. In an interview with the Maloof Foundation newsletter, *The Wooden Latch*, Slimen summed up his feelings about the move process:

Slimen Maloof, Sam and Alfreda's son, remembers the heart-wrenching night the first section of the house he grew up in moved: "It has been a bittersweet emotional roller coaster of an experience. To see a place you have known and loved your whole life, moving like boxcars, illuminated by streetlights in the middle of the night, was extremely eerie and surrealistic. Hundreds of thoughts flashed through my mind and the one I really remember was how glad I was that my mom didn't have to go through this."[31]

Typically, the building structure movers would arrive early to check the hydraulic jacks, the brakes, and the positioning of the building sections on the move vehicles lined up near the new Amethyst Gate at the existing site. White lights were strung horizontally along the sides of the historic building

---

31. Slimen Maloof as quoted in "Maloof Residence Relocation Completed!!," *The Wooden Latch,* Fall 2002.

> Opposite Top: Balcony/mezzanine (Section 3) at new site awaiting positioning
> Opposite Bottom: Tower room (Section 4) parked at new site
> Bottom: Tree house room (Section 3) up on hydraulic jacks at new site

elements. What ensued was a slow-moving, circus-like parade that took until dawn to reach the new site. Because the last portion of the route, Carnelian Street, was a steep hill, it took all night to travel the three miles. And the move team had to proceed slowly and carefully whenever the truck went over a curb or around a turn. Upon arrival at the new site the structures were left up in the air on the hydraulic jacks; the concrete foundations were poured in later days. After the historic building sections were safely installed at the new site, Bob Buettner would take the American team out for an early breakfast.

THE HAUL ROUTE
(HISTORIC STRUCTURES MOVE PATH)

As conceived by the design team and as shown in the construction drawings, the historic structures haul route was laid out as follows: After going over the curb from the existing site into the street, the move caravan would travel north up Amethyst Street until making a left turn on Lemon, then a right on Beryl, a left turn onto Hillside, a right on Carnelian Street, until a final right turn over the curb into the new site. Sam's four shop sections were moved successfully using this route. The deconstruction of the guesthouse and historic residence had already begun when the American structures movers realized that they were not going to be able to make the turn onto Hillside from Beryl Street with the guesthouse. The move team came to this conclusion after discovering that because the telephone company chose not to bury them, three overhead telephone trunk lines crossing the street at that intersection were too low to push up and over the traveling building elements. It was not possible to move the trunk lines because the utility relocation effort had been underway for the past two years, and the overhead line position was the telephone company's final arrangement.

Unfortunately for us, the only other east–west corridor wide enough to traverse with the building sections was Wilson, which was a closed private road between Beryl and Carnelian Streets. At that time Wilson was a Concordia Homes' multi-family housing development construction site.

In a true act of desperation, I drove over to tell our story to the Concordia Homes site project manager, Ron Coppers. After discussing the situation with his boss, Steve Topor, Ron said that Concordia Homes would move up the date for the final paving of Wilson in their development so that we could use it to get to Carnelian via Wilson with our move vehicles. The Concordia Homes building team completed the Wilson Street work eight months earlier than planned, and they didn't mind when we brought eight heavy trucks over their new road surface. Ron even built a little bridge for us over an arroyo. So the final historic structure move route was Amethyst to Lemon to Beryl to Wilson to Carnelian Street. Because of Ron and Steve's kindness, Sam and Alfreda's historic buildings were able to reach the new site, and the Sam and Alfreda Maloof Center for Arts and Crafts was able to open to the public on time.

WORKING IN THE RAIN

Back at the existing site, after the move of the main workshop structure to the new site, all that remained of Sam's shop was the foundation. Confident in his abilities as a woodworker, in 1954 Sam poured a twenty-foot-wide by forty-foot-long concrete slab; it was the first element of his shop, home, and carport. He worked outside on this slab for more than a year, putting his tools and furniture away at night in an old chicken coop. Finally some friends helped him frame out on the existing slab, the original twenty-by-forty-foot workshop, using lumber provided by Rugg Lumber in Upland.

Now forty-six years later, Sam worked in the open air again. Burge Construction set up a tent over the main shop's concrete slab for him. While occupying the tent, Sam and Larry White kept the Maloof furniture production going from November 1999 to May 2000, including during the unusually cold and wet winter of 1999–2000. David Wade worked in the

Maloof Woodworking craftsmen (left to right David Wade, Larry White and Mike Johnson)

finishing shop (Building J), which was a chilly place because the exterior redwood siding had been removed and reinstalled at the new finishing shop. Mike Johnson set up his shop in the unheated two-story wood barn building to the west of Building J.

As well as Sam's six-year furniture backlog, the shop had commissions for the Claremont Theological Seminary's Kresge Chapel walnut altar elements and the St. Maximilian Kolbe, Oak Park, California, walnut altar furniture and great sanctuary cross. All completed pieces were stored in the Building J east display room. Even the Maloof garage (Building N) was used for the production of the St. Maximilian Kolbe altar cross, which was so large that the woodworkers had to pull it out into the courtyard to turn it over. A list of some of the special pieces produced by the Maloof studio during the move years (1998 to 2001) includes the iconic No. 52–1998 "Freda Maloof Collection" maple, ebony, and purpleheart "Duet Music Stand No. 2"; the No. 2–1999 "Bill and Rosanna Baldwin" fiddleback maple and ebony rocking chair; the No. 23–"1st Prototype 24 May 1999" walnut and ebony rocking chair; the No. 3–2000 "Dr. and Mrs. Robert W. Edgar" walnut and ebony rocking chair; and the No. 12–2000 "Alfreda's Chair." Sam donated this special blonde maple rocking chair to the annual summer charity auction held by Anderson Ranch School of Woodworking in Snowmass, Colorado. After a spirited round of bidding, "Alfreda's Chair" sold for $120,000.

Some of the studio furniture produced by the Maloof team during the tent phase featured a sculptural edge that followed the natural outline of the tree. Described by woodworkers as a natural or free edge, this style provides the pieces with a rough-hewn, energetic look. Larry White told me that he thought Sam started using the free edge because it took less time to finish. So this looser, somewhat wilder Maloof furniture style could be a result of the brutal working conditions at the site at that time. For example, the St. Maximilian Kolbe altar is walnut with the free edge and exposed large dovetail joints; its tabernacle has a tree branch for a latch.

Sam was not the first woodworker to use this style of finishing; it has been used for centuries all over the world, such as for wooden Japanese temple screens. French architect Charlotte Perriand, an associate of Le Corbusier, was producing free-edge furniture in the 1930s. As early as 1941, Sam's friend George Nakashima also made a free-edge cabinet for Andre Ligne. And in 1986, at the age of eighty-one, Nakashima used the natural edge for the Cathedral of St. John the Divine's Peace Altar in New York City.

BUILDING REASSEMBLY STRATEGIES

While Sam and the boys concentrated on furniture production in the tent at the original site, the construction team prepared to reassemble the relocated buildings at the new site. During site preparation, trenches had been dug for the historic residence and woodworking shop foundation stem walls; gas, water, electrical, and septic utility lines were stubbed up. When the time came to reassemble the house, shop, and guesthouse at the new site, it was obvious that nothing was straight and true. For this reason, the relocated building sections were left up in the air on jacks and wooden cribbing until measurements could be taken for the foundation systems. After confirmation of the proper locations for the walls, plywood formwork with steel reinforcing bar (rebar) was installed and concrete was pumped into the trenches.

After removing the cribbing and steel move beams from the relocated structures, Brad Sutton and his American Heavy Moving and Rigging team used a system of hydraulic jacks to lower the building sections onto the foundation walls. When all the sections were down, the floor slabs were poured and the roofing material was replaced at all of the building separation joints. Then waterproofing was applied in these areas and around the skylights and clerestories.

Next, special walls were reconstructed, such as the concrete masonry unit (CMU) wall between the former Maloof office (Section 1) and the dining room (Section 2); the cyclopean wall on the west side of Larry and Mike's shop (Section B); and the historic cyclopean walls in the outer courtyard area. Bob Buettner and masonry foreman Jesse Mier used Bob's photographs to replicate the walls at the original site.

The original building systems were relocated and when necessary upgraded, including heat pumps, air-conditioning units, furnaces, electrical panels, and the shop dust collection system. For example, to meet code the electrical panel boxes in the shop were expanded from one to three boxes, and major new systems were installed, including fire sprinkler and burglar alarm systems.

After the Maloof doors and windows were reinstalled and the building separation joints sealed, the buildings were ready for the finishing phase of the work, such as the installation of new gypsum drywall, painting, and missing ceramic tile replacement. The previously removed design elements were also reinstalled, including the brick flooring, cast-iron wood-burning stoves, weather vanes, and two carousel horses. The school bell from the demolished garage (Building N) was reinstalled at the historic residence entry courtyard gate.

GUESTHOUSE REASSEMBLY

The relocated guesthouse was left up on cribbing until after the hydraulic jacks were removed, measurements were taken, and holes were dug for the foundation system. Because the guesthouse porch posts were resting on river rocks at the original site, the contractor also needed to create a more positive connection between the posts and rocks. For this reason, holes were drilled in the bottom of each post and rock. Then epoxy was used to embed anchors in the drilled hole. After the cribbing and steel move beams were removed, the guesthouse was lowered with jacks, and the posts set on the anchors in the rocks.

THE HOUSE FITS TOGETHER

After the furniture, art, and household furnishings were moved to the new site and installed close to their former positions in the historic Maloof residence, this special environment resembled the original site except that its function had changed from Maloof family home to museum and educational center. This major change in use resulted in Sam's move to the new residence and the Maloof office's relocation to the finishing shop's (Building J) east end.

NEW FINISHING SHOP (BUILDING J)

The finishing shop at the new site is a new building with new mechanical and electrical systems. It looks like the demolished Building J at the original site because the redwood siding was carefully removed and reinstalled on the new structure. More space was needed to move large pieces of furniture between the shops, so the new Building J was positioned several feet north of the original finishing shop's position at the old site.

NEW BUILDINGS N AND M

The new Maloof garage (Building N) and wood barn building (Building M) at the new site were included in the construction documents as an alternate, or option. The plan was to replace the garage (Building N) with a public restroom structure; only the plumbing was stubbed up and capped off. Several years later, the Maloof Foundation constructed the restroom structure in the Maloof-style with cyclopean walls.

HISTORIC CONTEXT RECREATION

The setting of the former Maloof homestead was overlaid onto the new site's scenic landscape; it was thus transformed into the Sam and Alfreda Maloof Center for the Arts and Crafts. For example, the school bell and tower from the demolished garage (Building N) were installed over the former home's (and now the museum's) entry courtyard gate. The large Japanese maple tree (*Acer palmatum*) next to the entry courtyard gate was successfully moved to the new site.

However, until that change was complete, the significant natural differences between the Carnelian and Highland Avenue properties made the reconstruction of the former historic setting difficult. This was an issue because a major goal of the project was to move the Maloof family compound in such a manner as to allow it to regain its place on the list of properties eligible for the National Register of Historic Places.

A significant portion of the new site was leveled to accommodate the historic building pads, except that the original natural grade was maintained near the top, or north, end of the Almond Street driveway. And to the northeast of the guesthouse, a coast live oak (*Quercus agrifolia*) sits on a mound about five feet tall, the top of which is at the native soil elevation. In relation to the guesthouse, the siting of this oak tree is close to that of Sam's avocado tree (*Persea americana* "Fuerte") at the original site.

North of the guesthouse one can see undisturbed soil above the retaining wall (children's wall) that runs east and west along the path from the visitor's parking area to the Almond Street driveway. The "children's wall" is so called because this wall is a perfect height for children to sit on, and Sam thought it would be a good place for teachers to line up their students on tour days.

Gary Moon insisted that to replicate the historic context, the new and relocated landscape elements were to be planted in the same positions held at the Maloof homestead. Because of the differing site conditions at the north end of the site, we received a waiver from this requirement for some of the plants, such as the Japanese maples along the drive. It was Moon's idea to plant the maples as a group on the north side of the guesthouse.

A coast live oak grove was created north of the children's wall with two trees moved from the Maloof family compound and several relocated oaks native to the new site. The tree movers boxed the oaks to be relocated while the trees were still in the ground, including the bottoms. All trees receiving this treatment were in shock for a time because their

roots were cut, so it was necessary for them to rest in place for several months before transport to the new site. In addition, some trees, such as oaks, could only be trimmed or moved during their dormant period, which occurs in the winter months.

When the time came to move the trees to the new site, cranes were used to lift the boxed trees out of the planting holes and onto flatbed trucks. At the new site, the trees were lifted into the new planting holes before the sides of the boxes were removed. The tree box bottoms were left to rot in place.

A seventy-foot tall Canary Island date palm (*Phoenix canariensis*) near the Highland gate was to be moved to the new site. However, because of the steep terrain at the Almond Street gate, the date palm and the jelly palm (*Butia capitata*), which was northeast of the guesthouse, were planted to the west of the new residence in a newly created palm grove. The tree moving company, La Cresta, explained that the Canary Island date palm could be relocated but it would need to be removed and replanted in one day.

Top: Date palm from SW of Highland gate arrives
Middle: Lifting of date palm off flat bed truck with crane
Bottom: Completion of date palm planting

On a cold but bright January morning they started work, and around noon the tree was lifted out of the hole and laid down in the driveway. Dolores and I started toward the tree just as the rats in the palm fronds hit the ground running. The tree movers tied up the fronds with rope; the rope would be allowed to disintegrate naturally to protect the tree's tender inner fronds. The weather had turned and it had begun to rain by the time the movers lifted the palm tree, with its six-foot-diameter root ball, onto a flatbed truck. Because the new site was at the top of a steep road, it was slow going for the truck with its heavy load. Upon arrival at the new site, the big palm was lifted off the truck with a large crane. Then, while the palm tree was swinging in the air, a lightning storm broke out. At this point there was no turning

Top: Date palm successfully transplanted to the west of new residence
Bottom: Placing of date palm root ball in a rain storm

back, so the La Cresta team proceeded to maneuver the tree over to the planting hole and to lower it into the ground, where it thrives today.

OLD BUT WELL-LOVED TREES

All the California native oaks were moved in the winter, their dormant season, and pruned the following summer. In some cases the tree canopies were raised to improve vehicular clearance and to protect the trees. The maintenance program included roots aeration with vertical mulching; the use of a layer of three- to four-inch coarse wood chips in the zone under the canopy; and slow, deep watering.

> Opposite: Large oak tree moves from near guesthouse to new site
> Bottom Left: Boxing the outer courtyard area olive trees in place
> Bottom Right: Lifting boxed olives with crane

A summary of the Maloof Tree Moving and Maintenance Schedule is included in the appendix.

California native oaks are drought-tolerant plants that, when over watered, are subject to root rot. The plan was for the zone under the oak tree canopies to be kept free of ground cover and thirsty plants such as citrus and palms. For this reason and because the relocated trees had their roots cut, the arborist installed Irrometers around the bases of the oaks, maples, and olives to closely monitor the root ball moisture content.

Key trees at the original site that were too old to move were replaced with same-size specimens, but not always with the same species. For example, the big California black walnut tree (*Juglans californica*) to the guesthouse's southeast was replaced with a 120-inch-boxed coast live oak (*Quercus agrifolia*) from the Maloof homestead south fence. When the oak tree was moved in January 2001, it weighed approximately 50,000 pounds on the crane (boxed).

However, the Irrometers failed to detect that the clayey soil in the new planting hole was holding in the water. It wasn't until the leaves started to drop that we knew something was wrong. Unfortunately, this tree did not survive. Because the oak tree died, it was replaced with a large Canary Island pine (*Pinus canariensis*), which seems to be tolerating the soil conditions.

Against the project arborists' advice, the four old, multi-trunked, twenty-five-foot-tall olive trees (*Oleo europaea*) in the Maloof parking/outer courtyard area west and south side planters were moved in ninety-six-inch boxes. As predicted, the trees struggled to survive at the new site, and had to be replaced. So with photos in hand, I selected at an olive farm near Bakersfield, California, four olives that matched the existing trees.

The California sycamore (*Platanus racemosa*) west of the woodworking shop was too large and old to move, so another seventy-two-inch-boxed sycamore replaced it at the new site. This tree is directly south of the Maloof studio woodworkers' Carnelian Street gate. Before the existing site demolition, Sam had a favorite tree he called his "Cedar of Lebanon" that was west of the Maloof Highland street gate and just north of the guesthouse. Actually a Deodar cedar (*Cedrus deodara*), it was replaced with a seventy-two-inch-boxed nursery-grown specimen that was brought by flatbed truck from Sacramento, California. The new Deodar cedar was planted northeast of the visitor's parking area in native soil. Sadly, a couple of years after project completion, a tremendous gust of wind coming down from the mountains threw the tree across the parking lot.

All that remains at the original site to mark the spot where the Maloof family compound stood is Sam's original "Cedar of Lebanon" and a pocket park created by the City of Rancho Cucamonga called the Alfreda Ward Maloof Commemorative Garden. Sam's tree can still be seen on the Route 210 freeway north embankment, just east of the Amethyst Bridge.

THE CEREMONY OF THE LATCHES

The *Docent Information for the Maloof Historic Home* manual includes the following account, presumably told by Sam: "When the house was moved, someone put all the hand-crafted historic Maloof residence door latches in a big box and Sam spent the day sorting them out."[32]

This comment is a good example of Sam's perceptions, and the impact that time can have on a story. Actually, so as not to overwhelm Sam, we didn't provide him with day-to-day project details, just a broad overview. Here is what really happened to the hinges-and-latches box:

Sanbag stationed guards 24/7 at the existing site because the historic residence itself was an art object. However, first a generator was stolen, and then a Maloof residence iconic latch. The missing latch resembled a flower; its image has been in numerous magazine articles, books, and videos. This latch was installed on the sitting room's south exterior double doors below the upper gallery stairs.

After the theft of those objects, Bob Buettner removed all the latches and hinges in the house. Then Bob documented the historic elements by photographing them on an illustration board. The objects were treated as art and wrapped by Cook's Crating. The latches and hinges spent the next year in a locked cage at the Graebel Crown warehouse along with the weather vanes.

After the Maloof property reconstruction was complete, Bob, Dolores, and I met Sam at the museum (the former historic Maloof residence) to give him back the house keys along with the latches

---

32. Various sources, "Stories About Sam," *Docent Information for Maloof Historic Home*, July 2012.

Iconic Maloof door latch, historic residence sitting room south door

box. The hinges had been reinstalled earlier, when the Maloof handcrafted doors were brought back from storage. We shared a moment as Sam looked in the box…and couldn't remember where all the latches went. After joining us in a hearty laugh, Sam led us from door to door during an impromptu Ceremony of the Latches. Sam just installed whatever latch seemed right. He went into his shop and made another latch for the door with the missing one.

MALOOF NEW RESIDENCE

After the historic Maloof residence was taken apart, reassembled, and repurposed as a museum and educational center, in June 2000 Sam reluctantly moved to the home he designed for himself and Alfreda. Alfreda had passed away in September 1998, and thus didn't live to see their new place except as an architectural model. The new residence can be seen today to the south of the former historic residence, on the northeast corner of Hidden Farm Road and Carnelian Street.

Sam accepted that he needed a place to live while his historic residence was being deconstructed. However, when the museum and educational center opened to the public, it was clear that he was not on board with the "museum" and "*former* Maloof home" concepts. Sam was upset by several changes to his former home that had been installed without his involvement, including the new exit signs, US American Disability Act (ADA) ramps, and a large red fire alarm panel behind his front door.

He brooded over the idea of visitors that he had not invited trooping through "his house." First Sam threw away the ADA ramps, and announced he was going to live in two places at once. After Sam made some Maloof-style changes to the new residence, he began to settle in there. The museum served as his guesthouse until Sanbag and the Maloof Foundation deemed that inappropriate. Maloof guests then slept at the new residence in the upstairs gallery or in the guest bedroom.

Sam lived alone in the new residence but he was not good at it. Then in July 2001 he married a special and talented person, longtime Maloof family friend Beverly Wingate. They first met in 1958 at the *California Design 6* exhibition held in the Millard Sheets Gallery at the Los Angeles County Fairgrounds. Awed by Sam's work, Beverly ordered a dining room table from him. More than forty years later, lightning struck. Beverly and Sam lived together in the new residence until the end of his life.

> Previous pages: New residence great room north window with Maloof-style redwood window treatment
>
> Right: Beverly Maloof writes in the new residence kitchen southwest sitting area

A wonderful master gardener, upon Maloof Relocation Project completion in September 2001, Beverly assumed garden management at the new site. Under Beverly's care the new site has become a beautiful California native demonstration garden. As the landscaping matures, the new site resembles the original site's historic context on Highland Avenue.

NEW RESIDENCE DESIGN

A superb example of Maloof architectural style, the new residence is a heavy timber-framed building sheathed in redwood with pitched blue sheet metal roofs and cyclopean walls. The main entry sequence to the new residence starts with a Japanese tori-style gate. It follows a landscaped path up to the hand-crafted front doors along the building's south side.

Upon being met graciously at the front entry, a visitor enters the light- and art-filled foyer. The entire home is a Maloof furniture- and art-filled experience. Straight ahead a Tony Abeyta painting hangs over a Larry White free-edge walnut credenza on which are many Mata Ortiz hand-thrown pots.

A Phillip Green lacquered mahogany canoe hangs overhead, near a gallery railing hung with Navajo blankets and rugs. (The artwork installation was not without its exciting moments. While we were installing the canoe, a fish line support snapped and it would have hit the brick floor if Sam hadn't caught it; he just laughed.)

North of the foyer is the heart and center of the structure, the great room, which is a dramatic three-story space with massive wood trusses, a second floor gallery, and clerestory windows. The great room provides magnificent mountain views as well as views of the museum and educational center on the hill above. The kitchen and garage wing are to the west of this space, and the bedrooms and office wing to the east.

A major issue with the new residence when first constructed was that upon entering the home through the entry foyer, a visitor would see the back of a straight run stair instead of a Maloof-style spiral staircase and the great room beyond. The stair completely blocked the view of the great room. Sam didn't understand why "you people" (a statement that included Sanbag personnel, the architects, the contractor, and me) couldn't see that a spiral staircase for the great room was the critical design element.

What Sam failed to appreciate was that the Sanbag budget for this house had been exceeded long ago. The funding earmarked for the new residence was for a simple caretaker's cottage, not a Maloof work of art. Gary Moon did appreciate the aesthetic need for the spiral staircase. The discussion was over forever when he said to me, "Sam will build it."

And Sam did. After the construction team finished their work and left the site, he started building again. Sam did not bond with the new residence until after the Maloof Relocation Project was finished, and he and his craftspeople were able to create another celebration of joinery with Maloof-style window and door trim, handcrafted doors, and an iconic handcrafted spiral staircase.

Sam and his boys also constructed a second entrance from the upper gallery, and another master bedroom with views of the mountains and Alfreda's burial plot. Sam and Beverly used the original bedroom as their office; they had Maloof handcrafted desks facing each other across the room. Sam's desk was the dining room table Beverly bought from him in 1958. Sam never did build the kitchen cabinets; that task remained unfulfilled. The young man who eventually built the kitchen cabinets gave this task his best effort, and his cabinets are still there.

The kitchen is a comfortable, light-filled rectangular space with colorful built-in sofas in one corner. A long walnut Maloof dining room table with built-in drawers along both sides fills the kitchen center area. Sam and Beverly started collecting the ceramics of Kevin Nguyen of the Xiem Clay Center in Altadena, California, and Nguyen's pots are arranged along the top of the dining room table.

Sometimes during construction Sam would invite a few of us down to the house to have a farm-style lunch with him at that table. Sam had a good story regarding this type of furniture: A customer called Sam to complain that a foul odor was emanating from her Maloof dining table. Sam visited the home and after walking around and around the table, discovered that a drawer was being used as a hiding place for vegetables.

When Sam first moved into the new residence, he brought with him a life-size white plaster nude sculpture of a woman. At the original site it was in the master bathroom. So we put the naked lady in Sam's bathroom at the new house, where she would startle people looking for the powder room. Beverly moved her to the pyramid room bathroom in the museum. Guests participating in historic home tours would peek at her from behind the door. She has since moved on to the museum's master bathroom.

SAM WANTS TO BE ALONE

Sam was not a solitary individual. He lived and worked with a core group of people, and he had friends and admirers all over the world. The historic Maloof residence and woodworking shop has always been a communal environment, one that is accessible and open to all, with almost no private areas. For example, Sam and Alfreda's bedroom in the historic residence was an art gallery and was always considered public space.

Upon moving to the new site, Sam felt the need for privacy. After he moved in to the new residence, folks were always going back and forth from the east porch through his bedroom. Sam said he needed a place where he could be alone when necessary. He wanted to build a Japanese teahouse constructed like fine piece furniture on a moon bridge over the arroyo. However, because the arroyo is a major storm drainage channel for water coming down from the mountains, the State of California said no. Sadly, this project remained unfulfilled.

MATA ORTIZ:
SAM GETS HIS GROOVE BACK

Sam's first major event at the new residence (and the new site) was the *Magic of Mata Ortiz* show held on February 17, 2001. This event was a reception honoring the master potters Jorge Quintana and Pilo Mora of Mata Ortiz in Chihuahua, Mexico. A percentage of the proceeds from the sale benefited the Mata Ortiz Children's Secondary School. The Mata Ortiz show was so successful that it has been held every year since at the new residence, during the weekend after Thanksgiving.

Sam was interviewed by *Metropolis* magazine just before his Valentine dinner with Beverly in 2007. It is a testament to his indomitable spirit that after all that had happened, he was able to say, "It's been fun. Oh yes, it's been beautiful." (Pronounced *bee-oo-ti-ful*.)[33]

---

[33] Chang, Jade. "The Craft Master: The Houses of Sam Maloof are Testaments to the Furniture Maker's Illustrious Half-Century-Long Career." *Metropolis* (April 18, 2007).

Top: New residence great room, Maloof rocking chair
Opposite Top: New residence dining room table and chairs with Kevin Nguyen pottery
Opposite Bottom: New residence loft detail, Native American blanket and Mata Ortiz pot

Opposite: New residence great room furniture grouping with double rocker

Top: New residence foyer, Larry White free-edge credenza

Bottom Right: New residence redwood spiral staircase detail

Bottom Left: Maloof free-edge pedestal table

EPILOGUE

Sam worked in his shop with his boys until about two or three months before his passing. Sam's body was weak but his old hands were steady and strong. Besides Larry, Mike, and David, Sam's last group of woodworkers included their apprentice, Isaac Bout, and Mike's son, Stephen Johnson. Two rocking chairs and a walnut rocking chaise lounge were some of the last pieces they produced together. Sam was able to find variations in his theme and maintain the integrity of his vision until the end.

Sam died at home on May 21, 2009, at age ninety-three. Some of his last words were, "Where am I going?" This was something that was often on Sam's mind during the Maloof Relocation Project. President Jimmy Carter called to say goodbye to his old friend on Sam's last day. So ended the life of a world-class woodworker and a generous person.

Business manager Ros Bock and the woodworkers continued the Maloof tradition in the relocated shop at the new site. The transition to life without Sam was difficult. However, after Sam's passing, the Maloof studio experienced a bump in furniture orders from the publicity generated by a major show that opened September 23, 2011, called *The House that Sam Built*, which was held in San Marino, California, at the Huntington Library, Art Collections, and Botanical Gardens.[34]

During the Maloof Relocation Project the studio maintained an approximately six-year order backlog, which means that a new customer could expect to wait six years for their furniture. Sam often joked that because it took so long for his clients to get their orders, the customers would tell him that they hoped to get their Maloof pieces before Sam passed on. Over time, the joke changed to receipt of Sam's furniture before the customer passed on. Maloof Woodworking's customer order backlog that began in 1953 finally came to an end in November 2013. At that time Larry, David, and Isaac left the shop to pursue their own interests. Slimen works in his shop at his ranch in Mentone, California. Ros Bock, Mike, and Stephen remain to carry on Sam's legacy.

The Maloof new site took its place on the National Register of Historic Places in 2010. By that time the historic residence and main shop were over fifty years old and the artist (Sam) had passed away. At a gala celebrating the property's register standing, President Jimmy Carter said, "You can take pride in the fact that the Maloof Relocation is one of the most significant preservation efforts in all of California history."[35]

Those of us working on the project came to know and understand Sam during the Maloof historic compound disassembly and reassembly. As the construction manager, it was my privilege to participate in the original structures' transformation to a museum and educational center that will allow future generations to be inspired by Sam's work. Project contractor Bob Buettner summed up the experience: "We were there to move Sam, and he moved us all."[36]

---

34. Harold B. Nelson, *The House that Sam Built: Sam Maloof and Art in The Pomona Valley, 1945–1985* (San Marino, CA: Huntington Library, Art Connections and Botanical Gardens, 2011).

35. President Jimmy Carter, "Celebration of Life Gala," Celebratory Remarks (video), Alta Loma, CA, June 4, 2011.

36. Robert Buettner, e-mail to Ann Kovara, January 2013.

Maloof Woodworking new showroom furniture display

APPENDIX

IN APPRECIATION

In appreciation for his remarkable insights, extraordinary support, and invaluable comments, I dedicate this book to my late father, Roy Barr. He was the first to read *Moving Sam Maloof*, which he read straight through with steady hands. He said, "It's good." So I kept going. It is one of my life's highlights that he was able to read the final draft just before he passed away at age ninety-six. Fortunately many other family members believed in the project as well, and provided their unwavering support: Simon and Ellie Kovara; Tom, Bev, and Mike Barr; Ellie, Greg, and Sarah Bertovich; Anne and Bill Groth; Kitty and Dave Martinez; Nancy Emily and Meagan Neal; and Anne Volk.

Special thanks to my friends for their support and sincere interest in my book. Their comments pointed me in the right direction: Angel Alvarez; Kiyo Bouhier; Winslow Bouhier; Bob DeLiso; Amanda Elioff; Elaine Ferguson; Barbara Gilliland; Laurene Harding and Luis Rivas; Sandi and Tom Keys; Ann and Randy Leach; Lucy Linnaus and Edgar Bennett; Susan MacAdams; Terry Marcellus; Vache Mardirosian; Deba and Kasturee Mohapatra; Dolores Moreira; Peggy Ollerhead; Phil Radler; Lisa Shepherd; Virginia Tanzmann; Rose Tourje; and Robert and Melva Young.

I owe a debt of gratitude as well to the folks at the Maloof new site for sharing with me their fond memories of Sam and Alfreda, their recollections of the Maloof Relocation Project details, and for assistance with this book: Beverly Maloof; Linda Apodaca; Ros Bock; Toni and Tom Bostick; Sam's good friend, Nick Brown; Evelyn George; Connie Ransom and her husband, Roger; Jim Rawitsch; John "Scotty" Scott; Dr. Joe Unis; and Maloof woodworkers Larry White, David Wade, and Mike Johnson.

Monumental thanks go to Bob Buettner for being a great contractor, the Maloof Project construction photographer, and a wonderful person. And thank you to the other *Moving Sam Maloof* photographers for contributing your amazing photographs: Ted Catanzaro, Kenny Kobrin, Tavo Olmos, and Norah Tahiri. Besides providing her beautiful photos, Norah incorporated the photographs and illustrations into the narrative and created the graphic design.

SAM AND ALFREDA: FURTHER READING

This is the sixth book published to date about Sam Maloof. It complements the others because it tells the previously untold Maloof Relocation Project story (1998 to 2001). My purpose in writing this is to express the beauty and inspiration my teammates and I experienced while moving Sam's place.

The first book to be published about Sam was his 1983 autobiography, *Sam Maloof, Woodworker*. Sam wrote a comprehensive description of his life and work from birth up to that time. *Sam Maloof, Woodworker* includes an introduction by Jonathan Fairbanks and photographs by Jonathan Pollock.

The second book about Sam, Jeremy Adamson's *The Furniture of Sam Maloof*, was published in conjunction with Sam's retrospective exhibition at the Renwick Gallery, Smithsonian American Art Museum, Washington, DC. Sam's Renwick retrospective was held from September 14, 2001, to January 20, 2002. Adamson's exhibition catalog provides an excellent overview of Sam's life and the Maloof Studio production up to that time. Sam worked closely with Adamson and photographer Jonathan Pollock on *The Furniture of Sam Maloof*.

Harold (Hal) B. Nelson's *The House that Sam Built: Sam Maloof and Art in the Pomona Valley, 1945–1985* was published next. This book was the exhibition catalog for the show held from September 24, 2011, to January 30, 2012, at the Huntington Library, Art Collections, and Botanical Gardens in San Marino, California. The Huntington exhibit featured key Maloof furniture pieces along with objects from the Maloofs' personal art collection and of their friends' and associates'.

The fourth book, Gene Sasse's *Maloof Beyond 90: An American Woodworker,* is a handmade, limited edition leather-bound large-format tribute to Sam. This special book features Sasse's photographs juxtaposed with essays by several authors. Sasse had the opportunity to present his unfinished work to Sam before Sam's death in 2009. Published in 2011, after Sam's passing, the fifth book is a smaller-format, paperback edition of that work, titled *Maloof at 90: An American Woodworker.*

Alfreda Maloof also wrote a book. Her autobiography, *Recollections from My Time in the Indian Service 1935–1943: Maria Martinez Makes Pottery,* was published in 1997. It describes her unique experiences living and teaching on several Indian reservations.

TECHNICAL RESOURCES, HISTORIC PROPERTIES

Because the Maloof site was declared a historic property by the State of California, it was under the jurisdiction of the historic section of the California Building Code. For this reason, we abided by the Secretary of the Interior's *Standards for Rehabilitation*.[37] One of the technical sources used for reference included the US Department of Housing and Urban Development's *Guidelines for Rehabilitating Old Buildings*.[38] A Historic American Building Survey (HABS) recording project for the Maloof existing site was conducted under the auspices of the US Department of the Interior's National Park Service division for the Library of Congress in Washington, DC.[39] The final product of the HABS survey included measured drawings, photographs, and a written descriptive narrative. Thirtieth Street Architects prepared the drawings, Tavo Olmos of Pasadena, California, was the photographer,[40] and Dr. Anthea Hartig, PhD, National Trust for Historic Preservation, prepared the narrative.[41] Robert Chattel, AIA, of Chattel Architects, Claremont, California, prepared the National Register of Historic Places application itself.[42]

PROJECT CONTRACTORS AND DESIGNERS

Sanbag used a two-part bid process for the selection of the Maloof Reconstruction Project general contractor. First a Request for Qualifications (RFQ) was posted and the responding contractors were pre-qualified for historic structure move experience, safety, and bonding capacity. The contractors deemed pre-qualified were invited to submit a bid. Sanbag made the final selection based on the lowest acceptable bid. The selected firm was Burge Construction. Led by project manager Bob Buettner, the Burge team brought to this complex effort previous experience

---

37. US Department of the Interior, *Secretary of the Interior's Standards for Rehabilitation and Guidelines for Rehabilitating Historic Buildings* (Washington, DC: Heritage Conservation and Recreation Service, January 1980).

38. US Department of the Interior, *Guidelines for Rehabilitating Old Buildings: Principles to Consider When Planning Rehabilitation and New Construction Projects in Older Neighborhoods* (Washington, DC: National Park Service, Office of Archeology and Historic Preservation, January 1977).

39. Harley J. McKee, *Recording Historic Buildings* (Washington, DC: US Department of the Interior, National Park Service, Historic American Building Service, 1970).

40. Tavo Olmos, "Sam and Alfreda Maloof Foundation for the Arts and Crafts—Photographic Record," US Department of the Interior, National Park Service, Historic American Building Survey (HABS), 1999.

41. Anthea Hartig, "Sam and Alfreda Maloof Foundation for the Arts and Crafts—Site Description," US Department of the Interior, National Park Service, Historic American Building Survey (HABS), 1999.

42. Robert Chattel, "Sam and Alfreda Maloof Foundation for the Arts and Crafts National Historic Site Nomination Form," National Register of Historic Places, 2009.

with successful historic preservation public works projects. And it was a good thing we had Bob, because he is a unique combination of competent contractor and sensitive guy. Besides having a complete grasp on the technical aspects of historic structure relocation, Bob understood the value of the Maloof environment, and the emotional issues surrounding the move for the Maloof family and craftspeople. As noted by former State of California Historic Preservation Officer (SHPO) and Maloof Foundation Director Dr. W. Knox Mellon, "Burge Construction did an excellent job of preserving the historic fabric and making the landscape and planters immediately surrounding the residence as close to the former setting as possible."[43]

Sanbag permitted Sam and Alfreda to select some design team members. The Maloofs chose Woody Dike to design the landscaping for the new site. Woody had worked in Sam's woodworking shop as a youth and was the son of their old friend, painter Phil Dike. Four decades earlier, Phil honored Sam and Alfreda by loaning them the remainder of the down payment money they needed for the purchase of the original Maloof family compound. So Woody Dike, ASLA, Ivy Landscape Architects Inc., Laguna Beach, California, was the project landscape architect. It was Woody who connected the Maloofs with Jim Wilson, the Maloof Relocation Project architect-of-record, who led the design team along with Carrie Wilde, AIA.

John Kariotis, PE, Kariotis and Associates Inc., Sierra Madre, California, was the structural engineer-of-record. Philip Douglas, PE, Associated Engineers Inc., Ontario, California, was the civil, utility, off-site, and traffic design engineer-of-record. Carolann Stoney, PE, Associated Engineers, Ontario, California, managed the multi-year utility relocation effort for the project. John DeSalvo, PE, Pacifica Engineers Inc., Laguna Hills, California, was the mechanical, electrical, and plumbing engineer-of-record. Ed Lyons, PE, RMS Group, Rancho Cucamonga, California, was the soils/geotechnical engineer-of-record.

SIX KINDS OF MOVERS AND THREE ARBORISTS

The relocation effort required six different types of movers and three arborists. Led by Brad Sutton, historic structure movers American Heavy Moving and Rigging, Chino, California, were subcontractors to Burge Construction. Atthowe Fine Arts Services packed and transported the Smithsonian-bound furniture, as did Action Air. Sanbag selected four other moving companies through a Request for Proposal (RFP) process based on qualifications and price. Cook's Crating of West Los Angeles, California, was selected to pack, move, store, and move again the art collection and Maloof studio handcrafted furniture. At the time they were the only movers qualified to move art for the Getty Museums in Los Angeles. The experienced crew at Graebel Van Lines, Ontario, California, packed and moved the smaller woodworking shop items, household goods, the Maloof office, Slimen's house, and the Maloof Studio's 200,000 board feet of lumber. Dan and Mike Kaplanek of La Cresta Landscaping successfully boxed, moved, and replanted the specimen trees and other landscape elements.

In September 1999, Badger Heavy Equipment Movers handled the Maloof studio heavy woodworking machinery transport. From the main shop, finishing shop, and Sam's driveway the movers relocated sanders, lathes, planers, table saws, drill presses, joiners, routers, shapers, a mortiser, and a dowling jig.

Because of its complexity, the Maloof tree relocation and landscaping work required three arborists' services. The project arborists included Ted Stamen, Certified Arborist and UC Riverside professor; Robert Nesbitt, Certified Arborist and Plant Pathologist; and Paul Pondella, Certified Arborist and President, Tree King Tree Service, Inc.

---

43. Dr. W. Knox Mellon as quoted in "Maloof Residence Relocation Completed!!," *The Wooden Latch,* Fall 2002.

FURNITURE, ART & LANDSCAPE INVENTORY

FURNITURE, ART & LANDSCAPE INVENTORY
MALOOF ART AND FURNITURE CATALOG (EXCERPT)

HISTORIC RESIDENCE LOCATION/OBJECT NO. – MEDIA – DESCRIPTION – CONDITION – SIZE – DESTINATION CODE – DATE MOVED – BOX NO.[44]
DR 108 – Chair, Sam Maloof, black walnut, tall back with horizontal horns, spindle back, low curved arms, bright orange leather seat, (Renwick piece #12 – Atthowe Movers picked up from Cook's warehouse on 12 April 2001), 37 X 21 X 19, (2), N/A.
DR 162 – Ceramic vessel, Peter Voulkos, tall greenish with specs & appliquéd elements and tall neck, 24 X 15, (1), moved 23 May 2000, Box 27.
GR 170 – Desk built-in, Sam Maloof, with log support on rock, black walnut, 29 X 54 X 25, (2), moved 6 July 2000, Box 705.
GR 56 – Ceramic bowl, "Maria and Julian [Martinez]," (San Ildefonso, 2.79) etched design diagonal & horizontal, 5 X 6-1/4, (1), moved 13 June 2000, Box 384.
TH 8 – Sculpture, wood carousel horse, unfinished with dog head on saddle, 30 X 60 X 12, (1), Moved 30 June 2000, Box 683.
MB 260 – Painting, watercolor Emil Kosa Jr., "The Fat Lady," 26-1/2 X 36-1/4, pale painted wood frame, 43-1/2 X 33-1/2, (1), moved 8 June 2000, Box 327.
MBA 1– Painting, Gerald Brummer, '96, "Chapel on Samos," white wood frame (28 X 38), 11 X 15, (1), moved 27 June 2000, Box 637.
M 96 – Ceramic bottle, Natzler tall tapered neck, striped side, base dark clay ochre with dark rim, 12-1/2 X 3-1/2, (1), moved 14 June 2000, Box 447.
DR 29 – Dining table, Sam Maloof, black walnut, eight drawers, two end leaves on wood hinges, 28-1/2H X 84 X 43, (1), moved 23 May 2000, N/A [Graebel movers blanket wrapped table].

The Maloof Catalog Destination code for this table is as follows:

1) Moved to new residence by Cook's Crating
2) Moved to Cook's warehouse
3) Moved by Graebel to new residence
4) Moved by Sam Maloof
5) Moved to Graebel / Crown Warehouse
6) Moved by Slimen Maloof
7) Moved attached to Building A-G

---

44. Amy Maloof and Katherine White, *Maloof Art and Furniture Moving Catalog,* June 12, 2001.

SMITHSONIAN-BOUND FURNITURE

LIST OF FURNITURE IN MALOOF HOME FOR RENWICK SHOW, APRIL 20, 2000[45]
1. "String" chair, ca. 1950 [unmarked] (unfinished) 29-½ X 24w X 33d
2. (Dreyfuss) prototype dining chair, 1952 [decal: 'Maloof,' gold on white] 29-¼ X 21w X 19-7/8 deep
3. (Raymond) dining chair, 1958 (yellow leather) ACTION AIR
4. (Kosa) occasional chair, 1954 [unmarked] (refinished) 37-¾ X 30-¼w X 28-¼d
5. ('Rejection') spindle-backed, occasional chair, sculpted arm, ca. 1955 (refinished) 34h X 29-¾ w X 32d
6. ('Radio') flared back, occasional chair, sculpted arm, ca. 1960 [brand] 39-½ X 27-¼w X 27d
7. (Raymond) dbl back, low arm, side chair, 1958 [brand] (refinished) 38 X 21-¾w X 25-1/8d
8. (Raymond) coffee table, 1958 [unmarked] (refinished) 17-½ X 33-½d X 54L
9. (Raymond) flared back, occasional chair, sculpted arms, 1958 [brand] 35-5/8 X 29-¼w X 30-¾d
10. (Raymond) dining chair, 1958 [unmarked; yellow leather] ACTION AIR
11. Stool, ca. 1958 [unmarked; yellow leather] (refinished) 29 X 34-½w X 21-¼d
12. Prototype horn back chair, ca. 1960 [unmarked; red leather] (cleaned) 37 X 21-5/8w X 21¼d
13. Evans chair prototype, low arm, [brand] '3-66' [black leather] (cleaned) 29-¾ X 20-¾ w X 21-½ d
14. Rocking chair, turned spindles, curved arm, '48 6-68 Alfreda Maloof' 44-7/8 X 26w X 42-1/8d. **Needs surface cleaning on one arm (ink?)
15. Rocking Chair, flat spindles, 'No. 68 1972 [sic] / Sam Maloof f.a.C.C. made for/ Alfreda L Maloof/ Christmas '72 [sic]/ P.S./ and all/ my love/ j.m.' 46h X 27w X 46-1/2d
16. (Aaron) child's rocking chair, turned spindles [unmarked] 28-¼ X 21-5/8w X 29d
17. Upholstered rocking chair, 1960s [brand] 44h X 27-1/2w X 39d
18. Upholstered settee, '8-67' [brand] 33-1/4h X 51-3/4w X 32d
19. 8-flat spindle occasional chair 'No. 49 1984/ Sam Maloof f.a.C.C/ j.m.' (to go with 21 & 22) 38 X 29-1/2w X 28d.. ***Reupholster with oatmeal???
20. Trestle table, 1960s [brand] (mezzanine) 25 X 20-1/2w X 73L
21. [Blank]
22. Game Table, 1960s, [brand] 28-¾ X 35-1/2w
23. Wing back chair (five spindles), 1960s [brand] 30-¾ X 27-7/8w X 20-1/8
24. Drop-leaf cabinet (bedroom), 1960s [brand] 34-1/2h X 53-¼ X 17-½ deep
25. Low–backed side chair (Woodenworks), 'No. 82 OCTOBER 1971 /RENWICK. SMITHSONIAN/ SAM MALOOF' 30-1/4h X 23-3/4w X 19-¼ deep
26. (Hlinka) musician's practice chair, 'No. 71 December 1972 For Jan Hlinka/ Sam Maloof/ and/ Sli-men Maloof' 29h X 24d X 18-3/4w
27. (Hlinka) double music stand, [unmarked; 1972] 44-½ X 51w X 30-1/2d

---

45. Jeremy Adamson, "List of Furniture in Maloof Home for Renwick Show," received Ann Kovara, Apr. 21, 2000.

SMITHSONIAN-BOUND FURNITURE (CONTINUED)

28. Trestle dining table, 'No. 2 1992 trestle made in 1975/Sam Maloof f.a.C.C./j.m.' (guest rm)
29. Spindle low back dining chair [brand; 9] 30h X 19-34w X 19-½ deep
30. ditto [brand; 5]
31. ditto [brand; 8]
32. ditto [brand; 1]
33. ditto [brand; 3]
34. ditto [brand; 6]
35. Desk hutch [unmarked, 1970s] 71-5/8 X 43-7/8w X 21-3/4w
36. Evans settee, 1960s [unmarked] (loft) 29-3/4h X 43-1/4L X 20-¾ deep
37. Spindle back settee, sculpted arm, 1960s (bottom of stairs) 38-½ X 45-1/2w X 22-¼ deep
38. Spindle back, low arm, dining/ side chair, [brand] '3 3/64' (purple upholstery) 39-5/8h X 21w X 19-¾ deep
39. (Slimen) Evans chair, 'No. 5 1987 Sam Maloof f.a.C.C.' 30-1/2h X 21-½ X 23-1/2d
40. Round pedestal table, 'No. 50/ 1997/'Sam Maloof No. 50/ 1997/ 'Sam Maloof No. 50 1997/ d.f.a. r.i.s.d./ m.j. l.w. d.w.' (bedroom) 28-1/2h X 32 dia
41. Low dresser (bedroom) 40-1/2h X 66-3/4L X 21-3/4w
42. Laced bench, 1960s [brand] (bedroom) 16-½ X 29-5/8w X 16-5/8d
43. Quartet music rack, 'No 42 1996 Sam Maloof d.f.a. r.i.s.d./ Alfreda Maloof Collection/ m.j. l.w. d.w." 49h X 30w
44. Double print rack, 'No. 17 1984/ Sam Maloof f.a.C.C./ j.m.' 38-¼ X 42w (fully extended) X 24-1/4d
45. (Mary Alice Frank) pedestal table [1988] 42 dia X 28-½
46. ½ rocker (guest room) 46-½ X 14w X 46
*45 pieces in all (5 Raymond) = 40 Maloof pieces; NB: #21

TREE MOVING & MAINTENANCE SCHEDULE

Descrip. of Plant / Box	Location at New Site	Location in Group	Irro-meters	Drip System	Location Existing Site	Date Trees Boxed	Moved to New Site	Set in Place	Sides Off
Weeping Bamboo/ 36"	Guest House Westside	East & West			Guest house westside	12/11/00			
Weeping Bamboo/ 60"	Guest House Westside	SW			Guest House Westside	12/11/00			
Deodar Cedar/ 72"	Visitor's Parking Area, Eastside			Yes	Nursery grown Sacramento				
Coast Live Oak/ 72"	Visitor's Parking Area, Eastside		Yes	Yes	San Marcos Nursery	N/A	5/7/01	5/7/01	5/7/01
Coast Live Oak/ 60"	Almond St Drive	West	Yes	Yes	San Marcos Nursery		6/11/01	6/11/01	6/11/01
Coast Live Oak/ 96"	Almond St Drive	Middle	Yes	Yes	Slimen's House Westside		3/13/01		3/14/01
Coast Live Oak/ 72"	Almond St Drive	NE	Yes	Yes	Amethyst Gate Eastside		3/14/01		3/23/01
Coast Live Oak/ 120"	Guest House SE		Yes	Yes	South Fence		1/8/01		3/13/01
Lg. Coast Live Oak	Guest House, Northeast on Mound				None - New Site Grown				
Sm. Coast Live Oak	Almond St Drive, Eastside				None - New Site Grown				
Sm. Coast Live Oak	Guest house, North of Children's Wall				None - New Site Grown				
Sm. Coast Live Oak	Guest House, Westside				None - New Site Grown				
Lg. Coast Live Oak	Carnelian St., North Gate				None - New Site Grown				

TREE MOVING & MAINTENANCE SCHEDULE (CONTINUED)

Descrip. of Plant / Box	Location at New Site	Location in Group	Irro-meters	Drip System	Location Existing Site	Date Trees Boxed	Moved to New Site	Set in Place	Sides Off
Japanese Maple/ 48"	Guest House, North side	West	Yes	Yes	Garage (Bldg. N), North side	12/14/00	1/09/01	1/17/01	3/12/01
Japanese Maple/ 24"	Guest House, North side	Middle	Yes	Yes	Highland Gate	12/14/00	1/09/01	1/17/01	3/12/01
Japanese Maple/ 48"	Guest House, North side	East	Yes	Yes	Garage (Bldg. N), North side	12/14/00	1/09/01	1/17/01	3/12/01
Phyllostachys nigra (Black Bamboo)/ 10' H X 10' W	Main Shop, Arbor Westside				Main Shop, Arbor Westside	12/14/00			3/12/01
Olive/ 26' W X 25' H, Multi/ 96"	Outer Courtyard, Southside		Yes	Yes	Outer Courtyard Southside	12/14/00	1/12/01	3/9/01	3/12/01
Olive/ 26'W X 25'H, Multi/ 96"	Outer Courtyard, Westside	Middle	Yes	Yes	Outer Courtyard, Westside	12/6/00	1/8/01	1/16/01	3/9/01
Olive/ 26' W X 25' H, Multi/ 96"	Outer Courtyard, Westside	South	Yes	Yes	Outer Courtyard, Westside	12/6/00	1/8/01	1/16/01	3/12/01
Olive/ 26' W X 25' H, Multi/ 96"	Outer Courtyard, Westside	North	Yes	Yes	Outer Courtyard, Westside	12/6/00	1/8/01	1/16/01	3/9/01
Japanese Maple/ 48"	Entry Courtyard Near Gate		Yes	Yes	Entry Courtyard Near Gate	12/6/00		2/23/01	
Saucer Magnolia/ 36"	Main Shop, Arbor Westside		Yes	Yes	Main Shop, Arbor Westside	12/14/00			3/12/01

TREE MOVING & MAINTENANCE SCHEDULE (CONTINUED)

Descrip. of Plant / Box	Location at New Site	Location in Group	Irro-meters	Drip System	Location Existing Site	Date Trees Boxed	Moved to New Site	Set in Place	Sides Off
Mission Fig/ 48"	Museum, Pyramid Rm., Eastside				Historic Residence, Pyramid Rm., Eastside				
Mission Fig/ 48"	Main Shop, Arbor Westside				Main Shop, Arbor Westside	12/14/00			3/12/01
CA Sycamore/ 72"	Carnelian St., South Gate			Yes	San Marcos Nursery	N/A	5/7/01	5/7/01	5/7/01
Lg. Coast Live Oak	Hidden Farm Gate				None - New Site Grown				
Canary Island palm/ 70' tall/ 6' Diameter Root Ball in Burlap	Lower Site/ Palm Grove				Guest House, Northeast	1/8/01	1/8/01	1/8/01	N/A
Jelly Palm/ 36" Diameter Root Ball in Burlap	Lower Site/ Palm Grove				Guest House, Northeast	1/8/01	1/12/01	1/8/01	N/A
Persimmon/ 48"	Lower Site/ West Fence				South Fence		1/12/01		3/13/01
Lg. CA Black Walnut	Lower Site/ West Fence				None - New Site Grown				

MAPS, SHOP DRAWINGS & FIELD NOTES

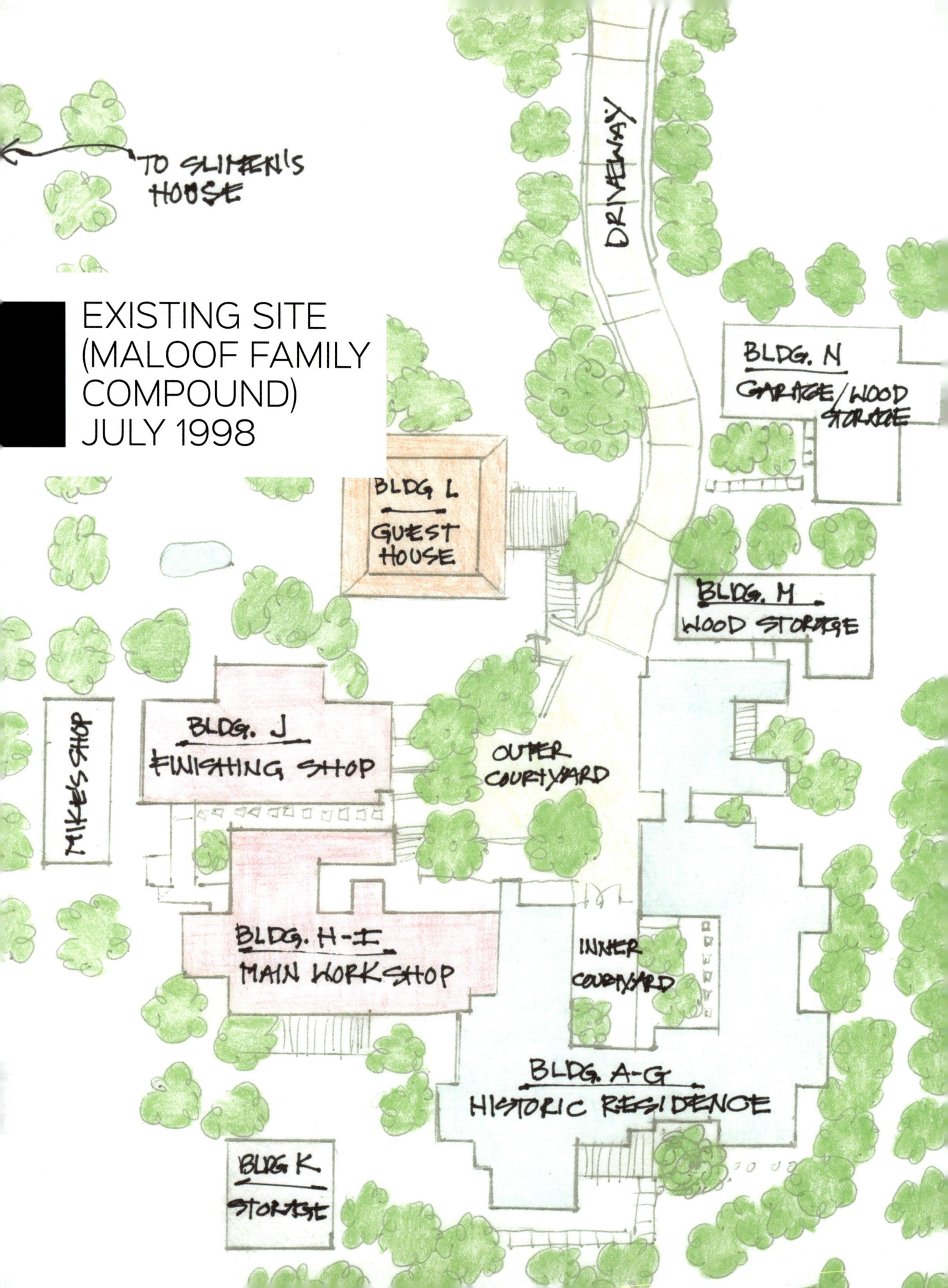

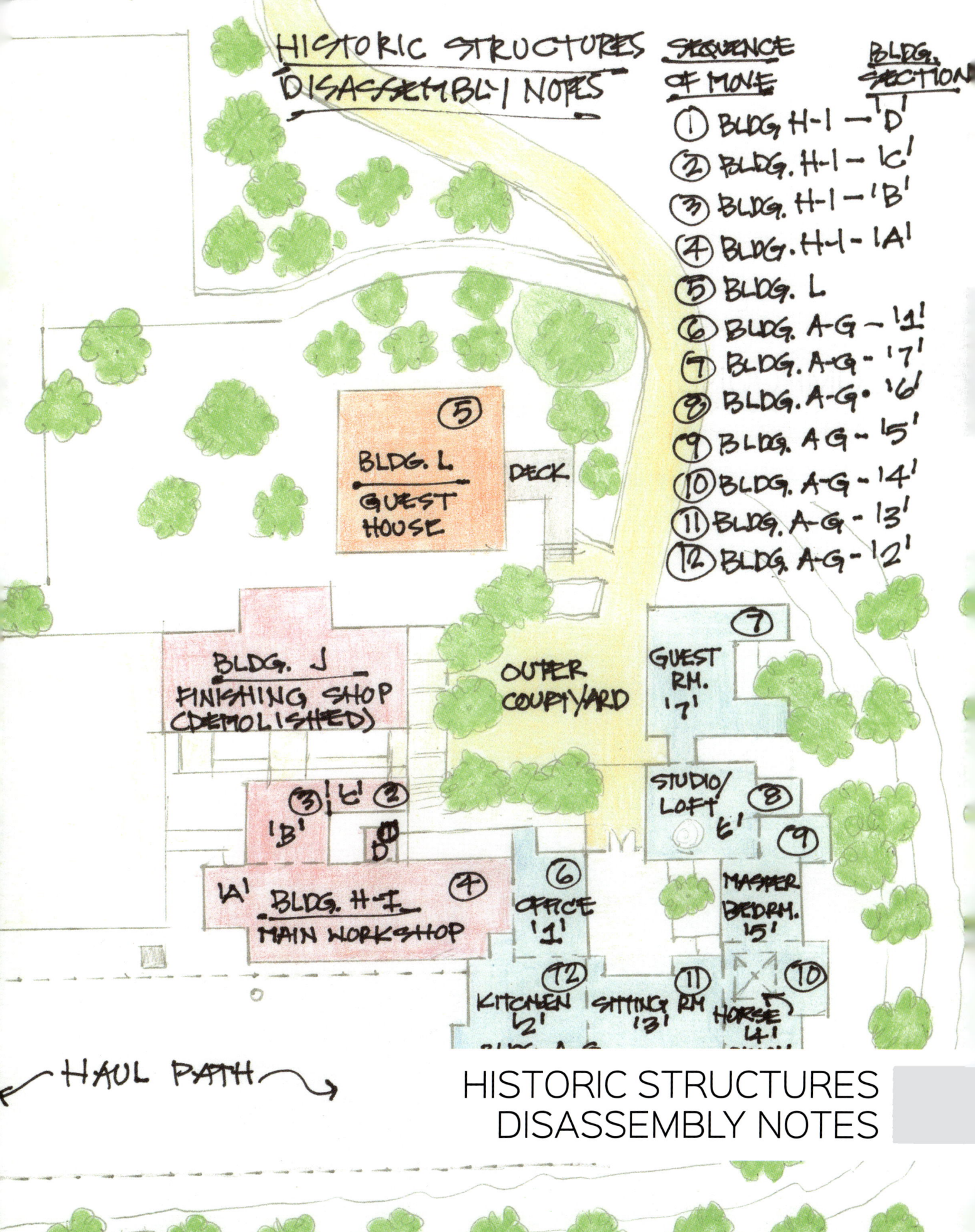

MALOOF STRUCTURES RELOCATION PLAN
NEW SITE

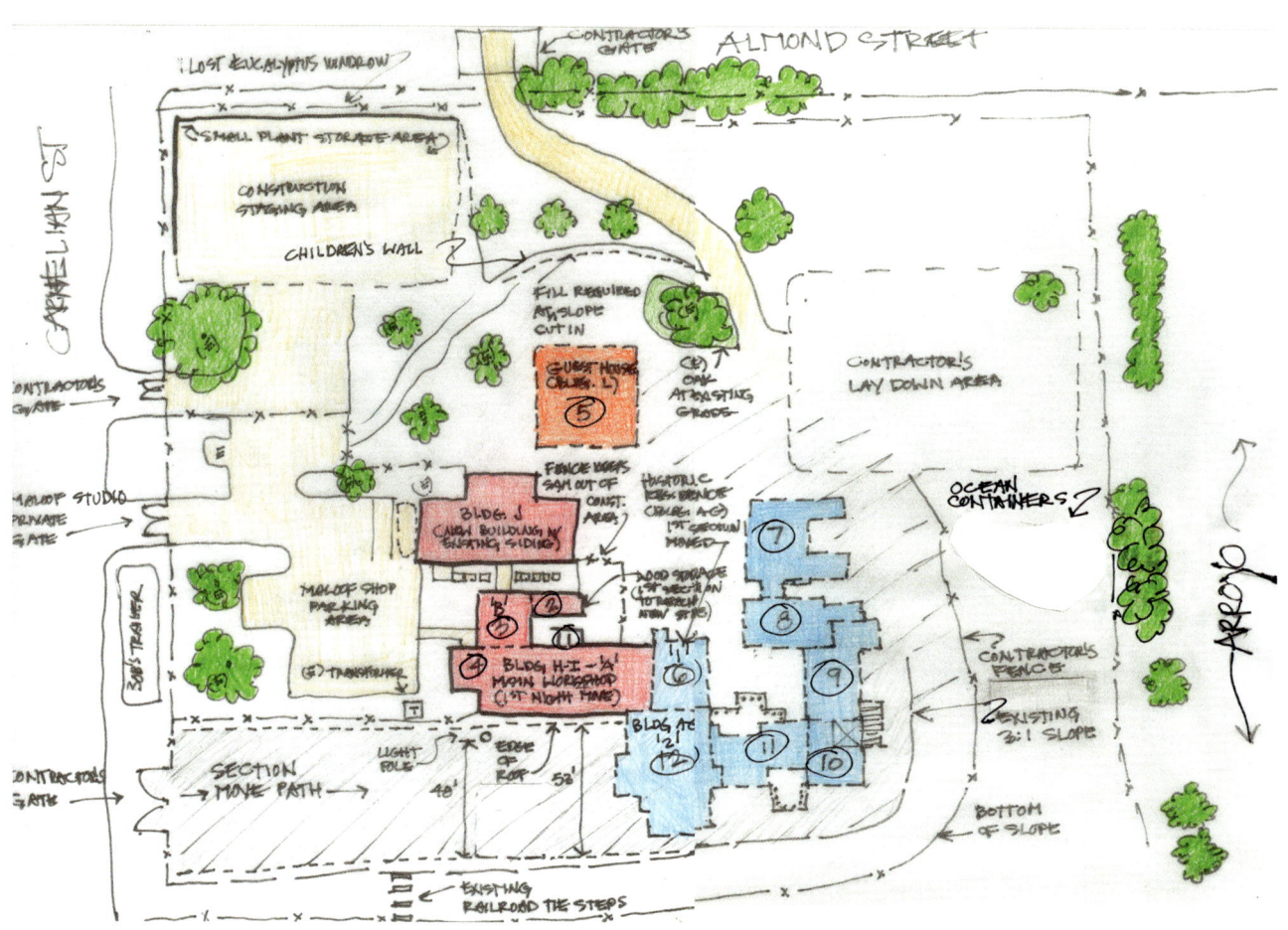

MAPS, SHOP DRAWINGS & FIELD NOTES 113

SEPARATION PLAN

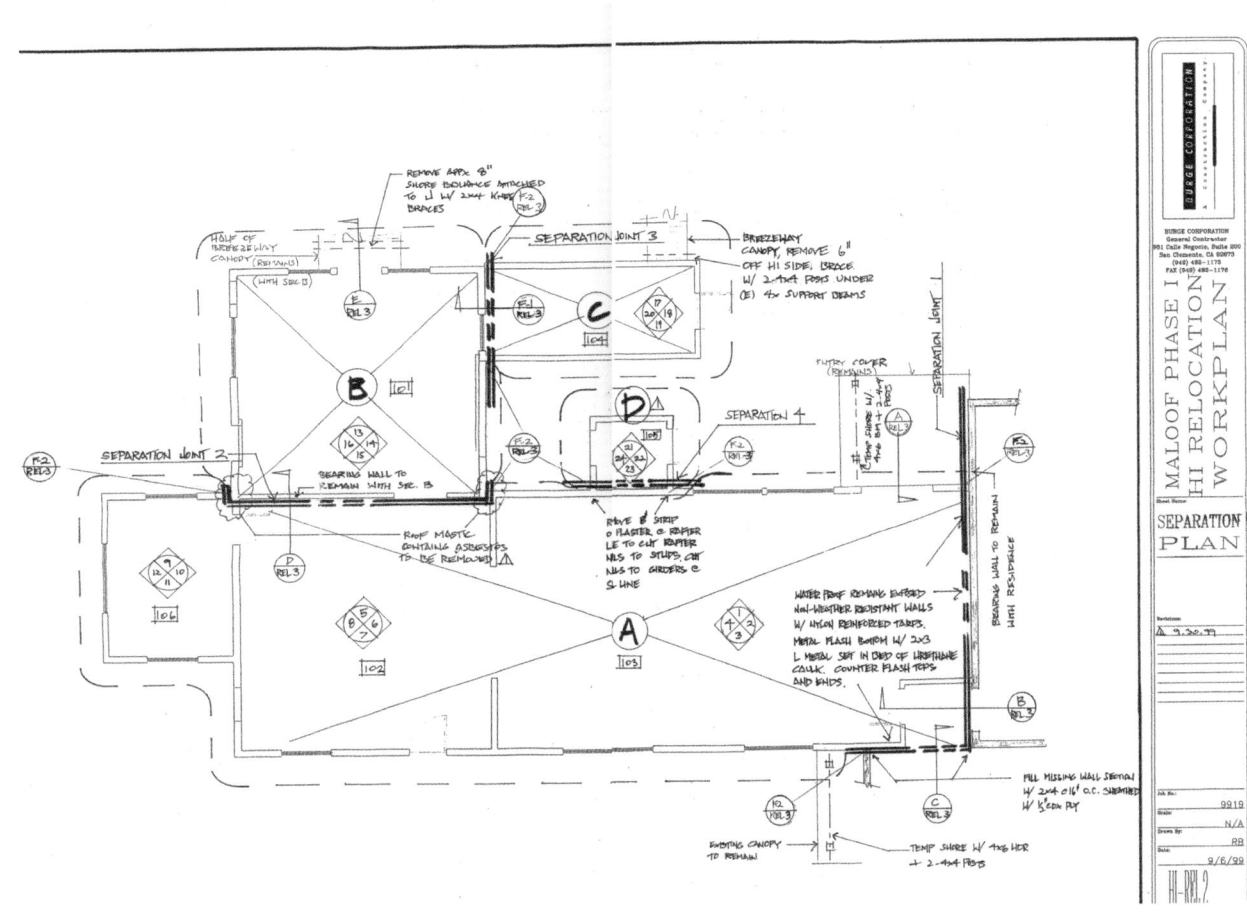

SEPARATION DETAILS

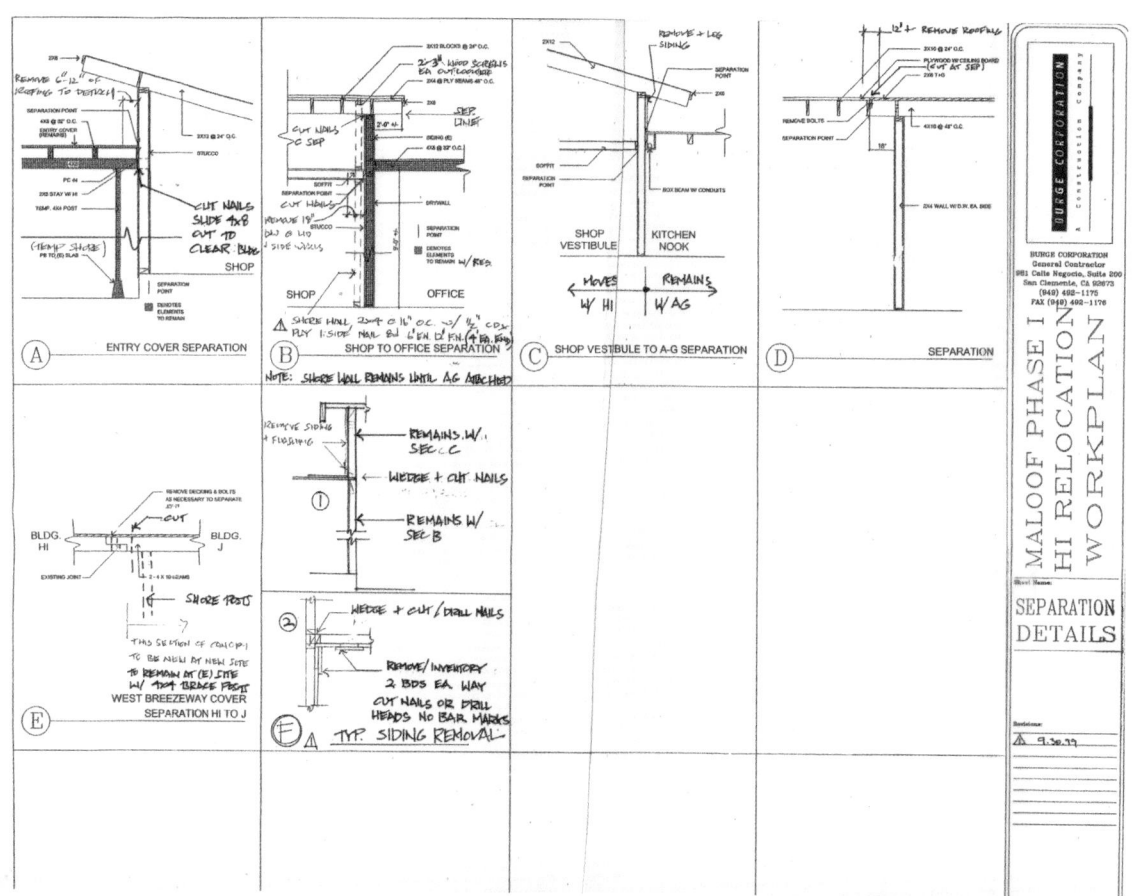

SHORE / LIFT PLAN

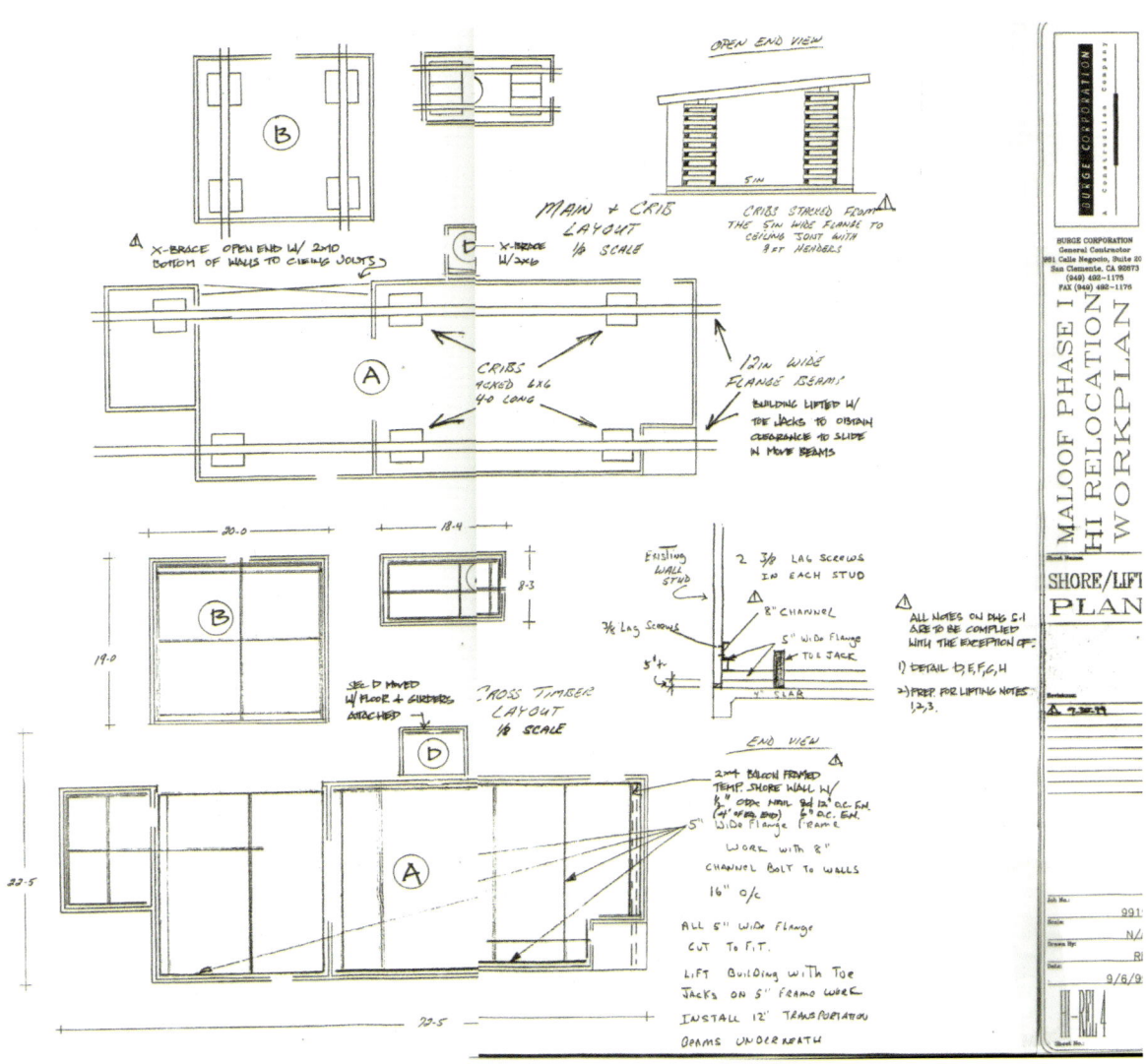

HAUL ROUTE

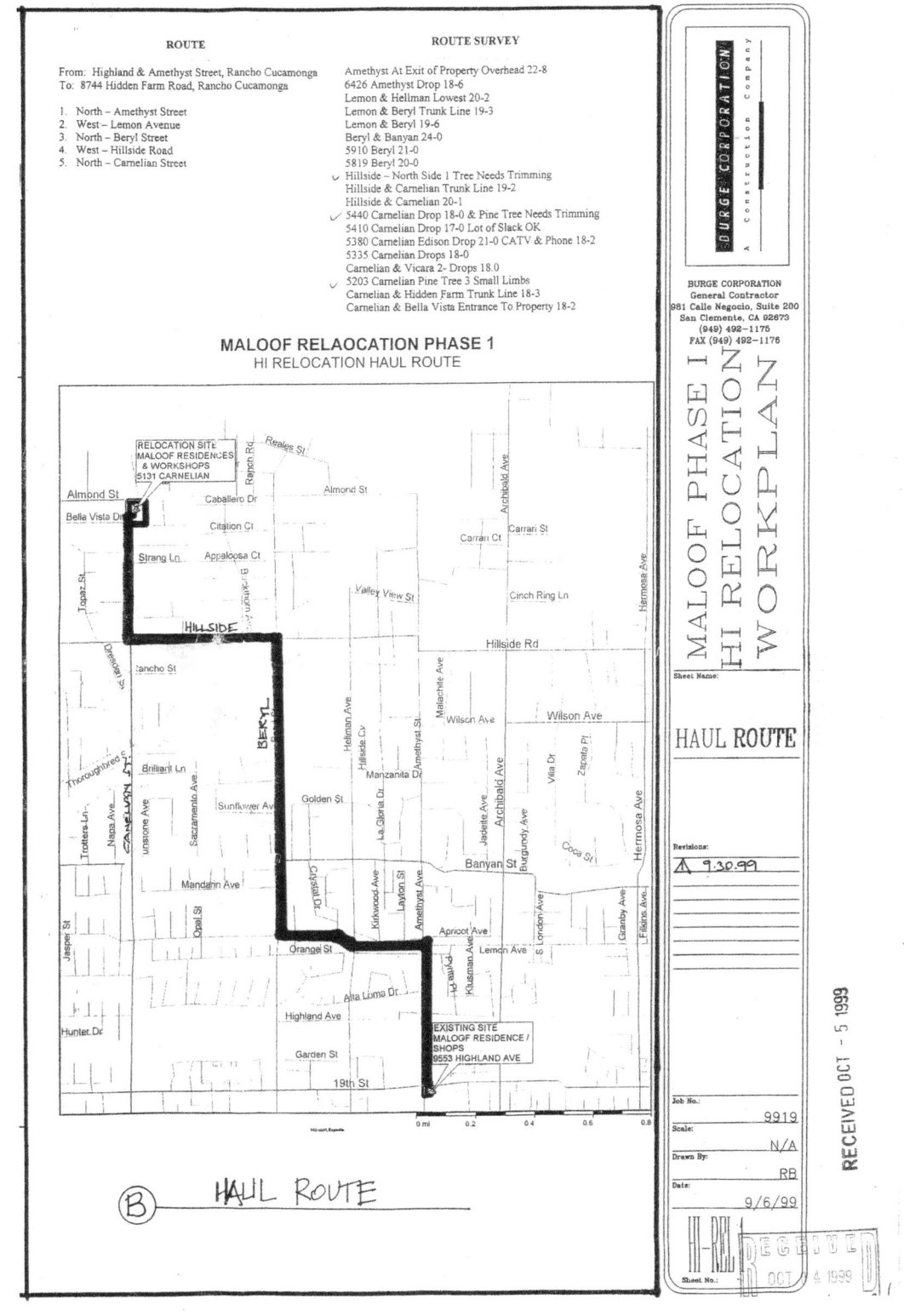

B — HAUL ROUTE

FIELD NOTES

Old Shop That is being Rebuilt At New Site

Removal of Redwood Door + Window Trim That has a DoveTail Detail At Joints And a Arched Head piece

Windows Are 3/4" Reveal

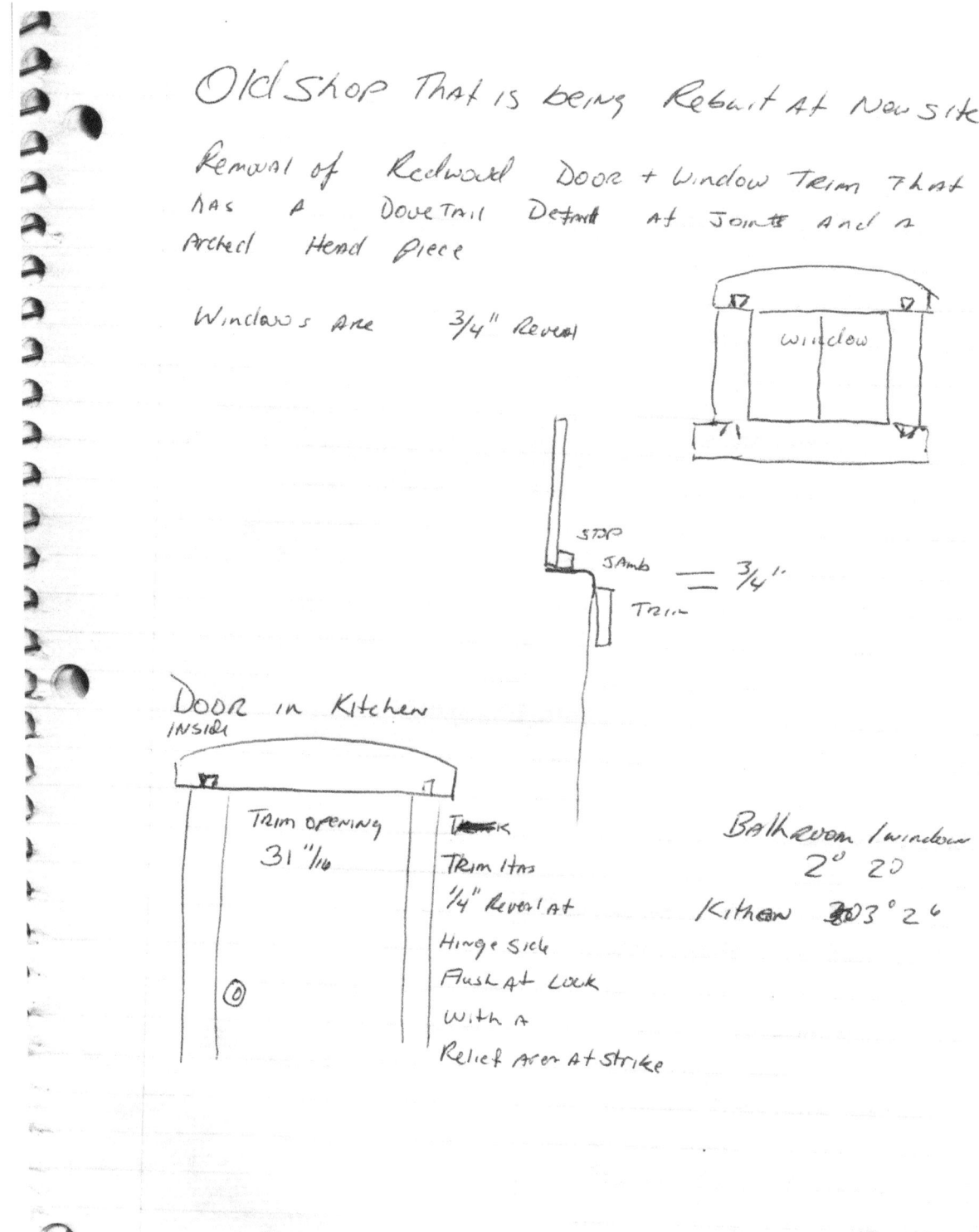

Door in Kitchen
INSIDE

Trim opening 31 11/16"

Trim Has 1/4" Reveal At Hinge Side Flush At Lock With A Relief Area At Strike

Bathroom / window 2° 20

Kitchen 3° 26

Door #2   Entrance to Kitchen
Redwood Trim on Inside   Head Flush with Jamb

                                                  Trim 1/4" Reveal
                        Trim Flush   At Hinge side
                        wit Jamb

Jamb width 31 7/16
Jamb Height 79 1/2
Door width 31 5/16
Strike From Head 43 1/2

Door #3   Kitchen side of Opening is Larger Than
Entrance to office Through Kitchen   office side of Door
Trim on Both sides of Door

Kithen side of Door Trim is Flush with Jamb All pieces

✱ Jamb width 32"
  Jamb HT   80 1/8

Door #3   OFFICE Door
Trim on the Inside of Office

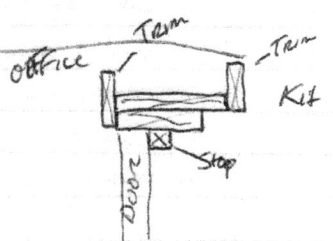

✱ Jamb width 30 7/16
  Jamb HT   79
  Door width 30 3/16   Trim Flush At Strike side
  Strike From Head 42 3/4   Trim 1/4" Reveal at Hinge side
                             Head is Flush

PHOTOGRAPHY CREDITS

ROBERT BUETTNER

    Pg 35, Top: Cyclopean wall construction
               Bottom: Cyclopean wall detail

    Pg 36, The last of Sam's avocado tree

    Pg 62, New site dedication, January 24, 2000 (Sam's 84th birthday)

    Pg 65, Historic residence, post-site demolition

    Pg 66, Stakes for historic residence at new site

    Pg 68, Section 2 disassembly, original site

    Pg 69, Top: Sections 2 & 3 on dollies
               Bottom Left: Balcony section (Section 3), NW corner
               Bottom Right: Section 3, NE corner

    Pg 72, Top: Balcony section (Section 3) at new site
               Bottom: Tower room section (Section 4)

    Pg 73, Section 3, tree house room, at new site

    Pg 74, Night shop move (Section A)

    Pg 75, Daylight move (Sections B, C, D)

    Pg 80, Top: Date palm from original site
               Middle: Lifting palm off truck
               Bottom: Shoveling sand on relocated palm

    Pg 81, Top: Date palm transplanted
               Bottom: Placing root ball of palm

    Pg 82, Oak moves from original to new site

    Pg 83, Bottom Left: Olives in outer courtyard
               Bottom Right: Boxed olives on crane

TED CATANZARO

    Pg 22, Alfreda Maloof portrait, 1997.

    Pg 23, Sam Maloof in orchard, 1997.

KENNETH KOBRIN

    Pg 13, New residence, great room

    Pg 88, Beverly Maloof in new residence kitchen

TAVO OLMOS

    Pg 2,   Sam Maloof, woodworker

    Pg 19, Redwood model in finishing shop (Building J)

    Pg 20, Dining alcove (Section 2), looking west

    Pg 32, Tower room (Section 4)

    Pg 37, Guesthouse, NE corner

    Pg 38, Top: Guesthouse (Building L), west garden
               Bottom: Interior guesthouse

    Pg 39, Garage (Building N), west elevation

    Pg 44, Top: Office (Section 1)
               Bottom: Balcony/mezzanine area (Section 3)

PHOTOGRAPHY CREDITS 121

Pg 45, Top: Upper gallery (Section 2)
Bottom: Master bedroom (Section 5)
Pg 46, Historic residence, spiral staircase (Section 6)
Pg 48, Top: Guest bath
Bottom: Pyramid room
Pg 49, Main shop (Section A)
Pg 52, Top: Guesthouse (Building L) north elevation
Bottom: Guesthouse interior elevation
Pg 53, Mike's shop (left) and David's shop (right)
Bottom: Main shop (Section B) west elevation

NORAH TAHIRI

Pg 8, Maloof rosewood double music stand
Pg 14, Main shop (Building A), chair patterns
Pg 21, Top: Main shop with furniture patterns
Bottom Left: Slabs of maple in wood barn
Bottom Right: New wood barn building
Pg 31, Looking south toward new residence
Pg 41, New wood barn building, Maloof Woodworking dining chairs
Pg 42, Historic residence entry courtyard gate
Pg 50, Maloof Woodworking: an Evans chair in progress
Pg 57, Relocated historic residence, east elevation
Pg 58, Maloof eucalyptus burl table
Pg 60, New residence great room, spiral staircase

Pg 61, Top Left: New residence, loft window
Top right: Sam Maloof picture frame
Bottom: Historic residence, Maloof dining table
Pg 77, Maloof Woodworking craftsmen, photos by Steve Scudder
Pg 85, Historic residence door latch, sitting room
Pg 86, New residence great room, north window
Pg 90, New residence great room, Maloof rocking chair
Pg 91, Top: New residence Maloof dining room table and chairs with Kevin Nguyen pottery
Bottom: New residence great room, Maloof rocking chair
Pg 92, New residence great room furniture grouping with Maloof double rocker
Pg 93, Top: New residence foyer, Larry White credenza
Bottom Right: New residence, spiral staircase detail
Bottom Left: Maloof free-edge table
Pg 95, Maloof Woodworking new showroom furniture display
Pg 128, Ann Kovara portrait

ILLUSTRATION CREDITS

MAPS

Kovara, Ann. Existing Site (Maloof Family Compound), July 1998

Kovara, Ann. Historic Structures Disassembly Notes

Kovara, Ann. Maloof Structures Relocation Plan, New Site

FIELD NOTES

Burge Construction. Maloof Phase I, Main Woodworking Shop Disassembly, *Dave's Dismantle Notes*, 2000

SHOP DRAWINGS

Buettner, Robert. Maloof Phase I, Main Woodworking Shop Relocation Work Plan, Separation Plan, 6 September 1999

Buettner, Robert. Maloof Phase I, Main Woodworking Shop Relocation Work Plan, Separation Details, 6 September 1999

Buettner, Robert. Maloof Phase I, Main Woodworking Shop Relocation Work Plan, Shore / Lift Plan, 6 September 1999

Buettner, Robert. Maloof Phase I, Main Woodworking Shop Relocation Work Plan, Haul Route, 6 September 1999

BIBLIOGRAPHY

Adamson, Jeremy. *The Furniture of Sam Maloof.* Exhibition Catalogue. Washington, DC: Smithsonian American Art Museum, Renwick Gallery, 2001.

Azar, George Baramki. "Soul of the Hardwood." *Aramco World* 46 (March–April 1995): 10–11.

Blake, William. *The Letters of William Blake (1906).* "Letter to the Reverend John Trusler," August 23, 1799.

Boardman, Allan. "Sam Maloof: Reflections of a Friend." *Fine Woodworking* 146 (Winter 2000/2001): 52–54.

Boggs, Brian. "Maloof's Challenging Chairs: A Fellow Chair Maker's Observations." *Fine Woodworking* 146 (Winter 2000/2001): 55.

Cathers, David. *Furniture of the American Arts and Crafts Movement.* Philmont, NY: Turn of the Century Editions, 1996.

Carter, President Jimmy. "Celebration of a Lifetime Gala." Celebratory remarks (video). Alta Loma, CA: June 4, 2011.

Chang, Jade. "The Craft Master: The Houses of Sam Maloof Are Testaments to the Furniture Maker's Illustrious Half-Century-Long Career." *Metropolis* (April 18, 2007).

Chattel, Robert. "Sam and Alfreda Maloof Foundation for the Arts and Crafts National Historic Site Nomination Form, National Register of Historic Places," 2009.

Curtis, John Obed. *Moving Historic Buildings.* Washington, DC: US Department of the Interior, Heritage Conservation and Recreation Service, Technical Preservation Services Division, 1979.

Diamonstein, Barbaralee. *Handmade in America: Conversations with Fourteen Craftmasters.* New York: Harry N. Abrams, 1983.

Fairbanks, Jonathan. "A Natural Devotion." *Antiques and Fine Art* 8 (May–June 1991): 65–69.

Green, Harvey. *Wood: Craft, Culture, History.* London: Viking Penguin, 2006.

Greene, Charles S. "Bungalows." *The Western Architect,* July 1908.

Hamilton, William. "Wood Works; For Life's Meaning, Examine the Grain." *New York Times*, September 13, 2001.

Harrington, Walt. *Acts of Creation: America's Finest Hand Craftsmen at Work.* United States: Sanger Group, April 11, 2014.

———. "An American Craftsman." *This Old House* (March/April 1998).

Hartig, Anthea. "Sam and Alfreda Maloof Foundation for the Arts and Crafts Site Description." Washington, DC: US Department of the Interior, National Park Service, Historic American Building Survey (HABS), 1999.

Hildreth, Jeremy. "Comfort and Joy: The House that Sam Built." *The Wall Street Journal*, January 12, 2012.

Jackson, J. B. *A Sense of Time, A Sense of Place.* New Haven, CT: Yale University Press, 1994.

Kammen, Michael. *Mystic Chords of Memory.* New York: Alfred A. Knopf, 1991.

Lang, Robert W. *Shop Drawings for Greene & Greene Furniture: 23 American Arts & Crafts Masterpieces.* East Petersburg, PA: Cambium Press Books, 2006.

Lauria, Jo, and Steve Fenton. *Craft in America: Celebrating Two Centuries of Artists and Objects.* New York: Clarkson Potter, 2007.

Makinson, Randell L. *Greene & Greene: Architecture as Fine Art.* Salt Lake City, UT: Peregrine Smith Books, 1977.

———. *Greene & Greene: The Passion and the Legacy.* Layton, UT: Gibbs Smith, 1998.

Maloof, Alfreda Ward. *Recollections of My Time in the Indian Service 1935–1943: Maria Martinez Makes Pottery.* Klamath River, CA: Living Gold Press, 1997.

Maloof, Amy, and Katherine White. "Maloof Art and Furniture Moving Catalog." June 12, 2000.

Maloof, Sam. *Sam Maloof, Woodworker.* Tokyo and New York: Kodansha International, 1983.

Mansfield, Howard. *Same Ax Twice: Restoration and Renewal in a Throwaway Age.* Hanover and London: University Press of New England, 2000.

Marsh, Lindell, Esq. "Maloof Property Chronology." Alta Loma, CA: Maloof Relocation Roundtable, September 23, 2011.

McCoy, Esther. *Modern California Houses: Case Study Houses, 1945–1962.* Los Angeles: Hennessey and Ingalls, 1977. First published 1962 by Reinhold.

McKee, Harley J. *Recording Historic Buildings.* Washington, DC: US Department of the Interior, National Park Service, Historic American Building Survey, 1970.

Nagyszalanczy, Sandor. "A Landmark Shop Destined to Become a Public Museum." *Setting Up Shop: The Practical Guide to Designing and Building Your Dream Shop*. Newtown, CT: Taunton Press, 2000.

Nakashima, George. *Soul of a Tree*. New York and Tokyo: Kodansha International, 1981.

Nakashima, Mira. *Nature, Form, and Spirit: The Life and Legacy of George Nakashima*. New York: Harry N. Abrams, 2003.

National Trust for Historic Preservation. "Historic Artists' Homes and Studios Program (HAHS): Sam Maloof Historic Residence and Woodworking Studio," 2003.

Nelson, Harold B. *The House that Sam Built: Sam Maloof and Art in the Pomona Valley, 1945–1985*. San Marino, CA: Huntington Library, Art Collections, and Botanical Gardens, 2011. Exhibition catalogue.

Noles, Pam. "Woodworker's Memories Splinter as Move Commences." Our Times sec., Inland Valley edition, *Los Angeles Times,* June 12, 2000.

Nuland, Sherwin B. *How We Die*. New York: Vintage Books, Random House, 1995.

Olmos, Tavo. "Sam and Alfreda Maloof Foundation for the Arts and Crafts Photographic Record." Washington, DC: US Department of the Interior, National Park Service, Historic American Building Survey (HABS), 1999.

Parks, Bonnie W., and Aaron A. Gallup. "Continuation Sheet 1, California Department of Transportation Architectural Form." February 17, 1989, rev. July 6, 1990.

Peart, Darrell. *Green & Greene: Design Elements for the Workshop*. Fresno, CA: Linden Publishing, 2005.

Quintana, Valerina. *This Piece of Earth: Images and Words from Tumamoc Hill*. Tucson, Arizona: Tumamoc: People and Habitats, College of Science, University of Arizona, 2013.

Riggio, Tod. "Maloof on the Move: Master Craftsman's Relocated Home to Become Cultural Center." *Workshop News* (Fall 2000).

Sam and Alfreda Maloof Foundation for the Arts and Crafts. "Maloof Residence Opens to Rave Reviews." *The Wooden Latch* (Winter 2003).

———. "Maloof Residence Relocation Completed!!" *The Wooden Latch* (Fall 2002).

———. "Maloof Residence Relocation Timeline." *The Wooden Latch* (Fall 2010).

———. "Maloof Site Nominated to National Register." *The Wooden Latch* (Summer 2003).

———. "Maloof Woodshop Thrives at New Site." *The Wooden Latch* (Summer 2003).

———. "Our Great Losses." *The Wooden Latch* (Fall 2009).

———. "Sam Maloof Residence and Woodworking Studio Joins [National Trust for Historic Preservation Historic Artists' Homes and Studios] Associate Sites." *The Wooden Latch* (Summer 2003).

———. "Sam Maloof, Woodworking Legend." Rene Russo, host. Los Angeles: KCET-TV PBS special, 2007.

Sasse, Gene. *Maloof at 90: An American Woodworker.* Alta Loma, CA: Gene Sasse, 2009.

Sauvion, Carol. "Craft in America: A Journey to the Origins, Artists, and Techniques of American Craft." Los Angeles: KCET-TV PBS special, originally aired May 30, 2007.

Sipchen, Robert. "A Man of the Woods." Life and Style sec., *Los Angeles Times*, July 24, 1994.

Skinner, Tina, and Steven P. Whitsitt. *Esherick, Maloof, and Nakashima: Homes of the Master Wood Artisans.* Atglen, PA: Schiffer Publishing, Ltd., 2009.

Stamen, Ted. "Arborist's Report for the Maloof Relocation Project." July 31, 1999.

"Taste Portrait of One Man Who Played the Game." *House Beautiful* (May 1954): 151.

Takes, Joanna Werch. "Class with a Woodworking Legend." (Photo of Alfreda's chair, No. 12, 2000, which sold for $120,000 at the Anderson Ranch auction.) *Woodworker's Journal* 24 (December 2000): 46–49.

Taunton Instructional DVDs. "Sam Maloof: A Fine Woodworking Profile." Newtown, CT: The Taunton Press, Inc., 2005.

Taylor, Joshua. Foreword to *Woodenworks: Furniture Objects by Five Contemporary Craftsmen.* St. Paul, MN: Minnesota Museum of Art with National Collection of Fine Arts, Smithsonian Institution, 1972. Exhibition catalogue.

Thirtieth Street Architects. "Sam Maloof Residence: Relocation Alternatives Analysis Report." Laguna Beach, CA: July 14, 1992.

Thompson, Diana. "Beneath the Surface: Restoration of the Japanese House." *Huntington Frontiers* (Spring /Summer 2012): 8–15.

Upshaw, Deborah. "An Interview with Sam Maloof." *Woodwork* (Summer 1989): 70.

US Department of the Interior. *Guidelines for Rehabilitating Old Buildings: Principles to Consider when Planning Rehabilitation and New Construction Projects in Older Neighborhoods.* Washington, DC: National Park Service Office of Archeology and Historic Preservation, January 1977.

———. "Historic American Buildings Survey: Sam and Alfreda Maloof Compound." San Francisco: National Park Service, October 16, 2000.

———. *The Preservation of Historic Architecture: The US Government's Official Guidelines for Preserving Historic Homes.* Washington, DC: Lyons Press.

———. *Respectful Rehabilitation.* Washington, DC: National Park Service Preservation Press, 1982.

———. *Secretary of the Interior's Standards for Rehabilitation and Guidelines for Rehabilitating Historic Buildings.* Washington, DC: Heritage Conservation and Recreation Service, January 1980.

Various. "Stories About Sam." *Docent Information for Maloof Historic Home*, July 2012.

Webster, John. "Handsome Furniture You Can Build." *Better Homes and Gardens* (March 1951): 258.

White, Larry. "Six to Sixty: Life at the Speed of Art—A Retrospective." *The Life Celebration of Katherine White*, March 2012.

Whitman, Walt. "2. Democratic Vistas." (1871), par. 118.

SELECTED WEBSITES
www.sammaloofwoodworker.com
www.malooffoundation.org
www.sam-maloof.com
http://www.en.wikipedia.org/wiki/sam_maloof
www.artistshomes.org

ANN KOVARA, AIA LEED AP, was the Maloof Relocation Project's on-site construction manager. A multi-talented architect, she serves as the California High Speed Rail Program project architect.
www.movingsammaloof.com